# Springer Optimization and Its Applications

VOLUME 97

**Aims and Scope**

Optimization has been expanding in all directions at an astonishing rate during the last few decades. New algorithmic and theoretical techniques have been developed, the diffusion into other disciplines has proceeded at a rapid pace, and our knowledge of all aspects of the field has grown even more profound. At the same time, one of the most striking trends in optimization is the constantly increasing emphasis on the interdisciplinary nature of the field. Optimization has been a basic tool in all areas of applied mathematics, engineering, medicine, economics, and other sciences.

The series *Springer Optimization and Its Applications* publishes undergraduate and graduate textbooks, monographs and state-of-the-art expository work that focus on algorithms for solving optimization problems and also study applications involving such problems. Some of the topics covered include nonlinear optimization (convex and nonconvex), network flow problems, stochastic optimization, optimal control, discrete optimization, multiobjective programming, description of software packages, approximation techniques and heuristic approaches.

More information about this series at http://www.springer.com/series/7393

Chongyang Liu • Zhaohua Gong

# Optimal Control of Switched Systems Arising in Fermentation Processes

清华大学出版社
TSINGHUA UNIVERSITY PRESS

Springer

Chongyang Liu
Zhaohua Gong
Mathematics and Information Science
Shandong Institute of Business
and Technology
Yantai, Shandong, China

ISSN 1931-6828 ISSN 1931-6836 (electronic)
ISBN 978-3-662-43792-6 ISBN 978-3-662-43793-3 (eBook)
DOI 10.1007/978-3-662-43793-3
Springer Heidelberg New York Dordrecht London

Jointly published with Tsinghua University Press, Beijing
ISBN: 978-7-302-37332-2 Tsinghua University Press, Beijing

Library of Congress Control Number: 2014949499

Mathematics Subject Classification: 49J15, 49J21, 65K10, 49M37, 92C42



Printed on acid-free paper

Springer is part of Springer Science+Business Media (www.springer.com)

# Preface

Switched systems have attracted much interest from the control community, not only because of their inherent complexity but also due to their practical importance with a wide range of applications in engineering, nature, and social sciences. Optimal control of switched systems, which requires determining both the optimal switching sequence and the optimal continuous input, has attracted many researchers recently. This phenomenon is due to the problem's significance in theory and applications. This book is not intended to compete with the many existing excellent books on optimal control theory and switched systems. We simply cannot write a better one! Our intention is to supplement them from the viewpoints of applications in fermentation processes.

The modern fermentation industry, which is largely a product of the twentieth century, is dominated by aerobic/anaerobic cultivations intended to make a range of high-value products. However, since most fermentation processes create very dilute and impure products, there is a great need to increase volumetric productivity and to increase the product concentration. As a result, significant work is needed to optimize the operation and design of bioreactors to make production more efficient and more economical. It is obvious that a model-based efficient approach is necessary to ensure maximum productivity with the lowest possible cost in fermentation processes, without requiring a human operator. Nevertheless, the mathematical determination of optimal control in a fermentation process can be very difficult and open-ended due to the presence of nonlinearities in process models, inequality constraints on process variables, and implicit process discontinuities.

In this book, we present some mathematical models arising in fermentation processes. They are in the form of nonlinear multistage system, switched autonomous system, time-dependent switched system, state-dependent switched system, multistage time-delay system, and switched time-delay system. On the basis of these dynamical systems, we consider the optimization problems including the optimal control problems and the optimal parameter selection problems. We discuss some important theories, such as existence of optimal controls and optimization algorithms for the optimization problems mentioned above.

The objective of this book is to present, in a systematic manner, the optimal controls under different mathematical models in fermentation processes. By bringing forward fresh novel methods and innovative tools, we are to provide a state-of-the-art and comprehensive systematic treatment of optimal control problems arising in fermentation processes. This can not only develop nonlinear dynamical system, optimal control theory, and optimization algorithms but also increase process productivity of product and serve as a reference for commercial fermentation processes.

## Acknowledgments

For the completion of the book, we are indebted to many distinguished individuals in our community. We would like to thank Prof. Enmin Feng and Prof. Zhilong Xiu, Dalian University of Technology, China, for bringing our attention to this area. Almost all the materials presented in this book are extracted from work done jointly with them. It is our pleasure to express our gratitude to Prof. Kok Lay Teo, Dr. Ryan Loxton, and Dr. Qun Lin, Curtin University, Australia, for their valuable comments during our visiting at Curtin University from January 2013 to July 2014.

We gratefully acknowledge the unreserved support, constructive comments, and fruitful discussions from Dr. Lei Wang, Dr. Yaqin Sun, and Dr. Qingrui Zhang, Dalian University of Technology, China; Dr. Jianxiong Ye, Fujian Normal University, China; Dr. Bangyu Shen, Huaiyin Normal University, China; and Dr. Jin'gang Zhai, Ludong University, China.

We are also grateful to Prof. Yuliang Han and Prof. Guang'ai Song, Shandong Institute of Business and Technology, China, for their kind invitations in publishing the book.

## Financial Support

We acknowledge the financial support from the National Natural Science Foundation of China under Grants 11201267, 11001153, and 11126077, from the Shandong Province Natural Science Foundation of China under Grant ZR2010AQ016, and from Shandong Institute of Business and Technology under Grant Y2012JQ02.

Yantai, Shandong, China
January 2014

Chongyang Liu
Zhaohua Gong

# Contents

# Chapter 1
# Introduction

## 1.1 Switched System

By a *switched system*, we mean a hybrid dynamical system consisting of a family of continuous-time subsystems and a rule that orchestrates the switching between them [123]. Many systems encountered in practice exhibit switching between several subsystems depending on various environmental factors [63, 262, 281]. Another source of motivation for studying switched systems comes from the rapidly developing area of switching control. Control techniques based on switching between different controllers have been applied extensively in recent years, where they have been shown to improve control performance [100, 128, 181]. Switched systems have numerous applications in the control of mechanical systems, automotive industry, aircraft and air traffic control, switching power converters, and many other fields.

The switching rules in switched systems can be classified into *state-dependent* versus *time-dependent switching* and *autonomous* versus *controlled switching* [59]. For a state-dependent switching, we suppose that the continuous state space (e.g., $\mathbb{R}^n$) is partitioned into a finite or infinite number of operation regions by means of a family of *switching surfaces*, or *guards*. In each of these regions, a continuous-time dynamical system (described by differential equations, with or without controls) is given. Whenever the system trajectory hits a switching surface, the continuous state jumps instantaneously to a new value, specified by a *reset map*. In contrast, for a time-dependent switching, the continuous-time dynamical system's switchings are activated according to time functions, i.e., a switching occurs at a certain time instant. These switching instants can be prescribed a priori and fixed or designed arbitrarily by engineers. On the other hand, by autonomous switching, we mean a situation where we have no direct control over the switching mechanism that triggers the discrete events. This category includes systems with state-dependent switching in which locations of the switching surfaces are predetermined as well as systems with time-dependent switching in which the rule that defines the switching

 
C. Liu, Z. Gong, *Optimal Control of Switched Systems Arising in Fermentation Processes*, Springer Optimization and Its Applications 97,
DOI 10.1007/978-3-662-43793-3_1

signal is unknown (or was ignored at the modeling stage). In contrast with the autonomous switching, in many situations the switching is actually imposed by the designer in order to achieve a desired behavior of the system. In this case, we have direct control over the switching mechanism (which can be state-dependent or time-dependent) and may adjust it as the system evolves. For various reasons, it may be natural to apply discrete control actions, which leads to systems with controlled switching.

As a special class of hybrid systems, switched systems are inherently nonlinear and non-smooth, and therefore many of the results available from the vast literature on linear systems and smooth nonlinear systems do not apply. Consequently, many basic system theoretic problems like well-posedness, stability, controllability, observability, safety, etc., and many design methods for controllers have to be reconsidered within the hybrid context. A system is said to be *well posed* if a solution of the system exists and is unique given an initial condition (and possibly input signals) [67]. The well-posedness property indicates that the system does not exhibit deadlock behavior (no solutions from certain initial conditions) and that determinism (uniqueness of solutions) is satisfied. The basic problems of stability for switched systems were discussed in [134]. Then, various methods have been developed to analyze stability through various types of Lyapunov functions such as common Lyapunov function [59], multiple Lyapunov function [35], surface Lyapunov function [89], etc. The other stability results of switched systems are presented in [64, 106, 137, 178, 277]. For the controllability concept and its historical comments, one may refer to [232] and references therein. The complexity of characterizing controllability and stabilizability has been studied in [33]. Controllability problem for piecewise linear systems has been studied; see, for example, [76, 86, 130, 269]. A similar story holds for observability and detectability [11, 22, 56]. For switched systems, a wide body of literature exists on the development of stabilizing controllers [178, 261] and model predictive control [25, 176, 182]. In this book, we shall focus on the optimal control of switched systems arising in fermentation processes.

## 1.2 Optimal Control

*Optimal control problem* is to determine the control policy that will extremize (maximize or minimize) a specific performance criterion, subject to the constraints imposed by the physical nature of the problem. Over the years, optimal control theory has been applied to a diverse collection of problems [38, 114, 205].

### *1.2.1 Standard Optimal Control*

Optimal control theory is an outcome of the calculus of variations, with a history stretching back over 300 years [216]. In 1638, G. Galileo posed two shape problems:

the shape of a heavy chain suspended between two points (the catenary) and the shape of a wire such that a bead sliding along it under gravity traverses the distance between its endpoints in minimum time (the brachistochrone). Later, L. Euler formulated the problem in general terms as one of finding the curve $x(t)$ over the interval $a \leqslant t \leqslant b$, with given values $x(a)$, $x(b)$, which minimizes

$$J = \int_a^b L(t, x(t), \dot{x}(t)) \mathrm{d}t \tag{1.1}$$

for some given function $L(t, x, \dot{x})$, where $\dot{x} := \mathrm{d}x/\mathrm{d}t$, and he gave a necessary condition of optimality for the curve $x(\cdot)$

$$\frac{\mathrm{d}}{\mathrm{d}x} L_{\dot{x}}(t, x(t), \dot{x}(t)) = L_x(t, x(t), \dot{x}(t)) \tag{1.2}$$

where the suffix $x$ or $\dot{x}$ implies the partial derivative with respect to $x$ or $\dot{x}$. In a letter to Euler in 1755, J.L. Lagrange described an analytical approach, based on perturbations or "variations" of the optimal curve and using his "undetermined multipliers," which led directly to Euler's necessary condition, now known as the "Euler-Lagrange equation." Euler enthusiastically adopted this approach and renamed the subject "the calculus of variations."

However, modern optimal control theory was established in the late 1950s since R. Bellman introduced *dynamic programming* to solve discrete-time optimal control problems [21], L.S. Pontryagin developed *minimum principle* [202], and R.E. Kalman provided linear quadratic regulator and linear quadratic Gaussian theory to design optimal feedback controls [113]. Subsequently, the existence of the optimal control for optimal control problems was widely investigated [41, 42, 78, 214, 233]. The optimality conditions were also discussed in [51, 69, 154, 155, 201].

Some optimization problems involve optimal control problems, which are considerably complex and involve a dynamic system. There are very few real-world optimal control problems that lend themselves to analytical solutions. As a result, using numerical algorithms to solve the optimal control problems becomes a common approach that has attracted attention of many researchers, engineers, and managers. The numerical solution of the optimal control problems can be categorized into two different approaches: (1) the direct and (2) the indirect method [236]. Direct methods are based on discretization of state and/or control variables over time and then solving the resulting problem using a nonlinear programming solver. Based on the discretization of the state and/or control, direct methods can be categorized into three different approaches. The first approach is based on state and control variable parameterization [73, 74, 77, 212, 250]. The second approach is control parameterization [101, 127, 139, 156, 211, 240]. The third approach is based on state parameterization only [107, 230]. Indirect method solves the optimal control problem by deriving the necessary conditions based on Pontryagin's minimum principle. The first step of this method is to formulate an appropriate two-point boundary value problem (TPBVP), and the second step is to solve the TPBVP numerically [37, 116, 177, 190].

For a dynamic system in the optimal control problem, a system which is governed by a set of ordinary differential equations is called *lumped parameter system*. In contrast, if a system is governed by a set of partial differential equations, then the system is called a *distributed parameter system*. In this book, we shall only deal with optimal control problems involving lumped parameter systems. For the optimal control of distributed parameter systems, we refer the interested reader to [1, 39, 58, 79, 142, 244] for details.

### 1.2.2 Optimal Switching Control

For optimal control problem of switched systems, the added flexibility of being able to switch between subsystems greatly increases the complexity of searching for an optimal control. In the most general case, determining an optimal control strategy for a switched system involves determining an optimal continuous input function and an optimal switching sequence.

The problem of determining optimal control laws for switched systems has been widely investigated in the last years, both from theoretical and from computational points of view [274]. The available theoretical results usually extend the classical minimum principle or the dynamic programming approach to switched systems. For continuous-time hybrid systems, general necessary conditions for the existence of optimal control laws were discussed in [36] by using dynamic programming. Necessary and/or sufficient optimality conditions for a trajectory of a hybrid system with a fixed sequence of finite length were derived using the minimum principle in [71, 199, 223, 238]. The existence of optimal control for switched systems was investigated [68, 221, 279]. The computational results take advantage of efficient nonlinear optimization techniques and high-speed computers to develop efficient numerical methods for the optimal control of switched systems. The problem of optimal control of switched autonomous systems was studied for a quadratic cost functional on an infinite horizon and a fixed number of switches in [87, 220]. Gradient-based algorithms for solving the switching instants in switched autonomous systems were developed in [72, 160, 273]. A two-stage optimization methodology was proposed for optimal control of switched systems with control input [272, 275]. Based on a parameterization of the switching instants, an optimal control approach was developed in [133, 158]. Essentially different from the results mentioned above, the switched system was embedded into a larger family of nonlinear systems that can be handled directly by classical control theory [26–28]. By adopting such problem transformation, there is no need to make any assumptions about the number of switches nor about the mode sequence at the beginning of the optimization. The possible numerical nonlinear programming technique under this framework was explored in [259]. It showed that sequential quadratic programming can be utilized to reduce the computational complexity introduced by mixed integer programming. The effectiveness of the proposed approach was demonstrated through several examples. Recently, the problem of computing the schedule of modes in switched systems was investigated in [9, 40, 138, 224, 258].

The vast majority of optimization techniques for switched systems, including those mentioned above, are restricted to switched systems without time delays. However, time delays are common in practical engineering systems [208]. Indeed, switched systems with time delays have various applications in areas such as power systems [175] and network control systems [121]. The presence of delays in a switched system complicates the search for an optimal control policy. Necessary conditions for determining optimal switching times and/or optimal impulse magnitudes for such systems were derived in [66, 248, 249] via classical variational techniques. Based on a parameterization scheme in which the switching instants are expressed in terms of the subsystem durations, an effective optimal control algorithm for switched autonomous systems with single time delay was presented in [268].

## 1.3 Fermentation Process

Fermentation is a very ancient practice indeed, dating back several millennia. More recently, fermentation processes have been developed for the manufacture of a vast range of materials from chemically simple feedstocks right up to highly complex protein structures.

### *1.3.1 Generic Fermentation Process*

The origins of fermentation are lost in ancient history, perhaps even in prehistory. However, "fermentation" has many different and distinct meanings for differing groups of individuals. In the present context, we intend it to mean the use of selected strains of microorganisms and plant or animal cells for the manufacture of some useful products or to gain insights into the physiology of these cell types [170]. By contrast, the modern fermentation industry, which is largely a product of the twentieth century, is dominated by aerobic/anaerobic cultivations intended to make a range of high-value products.

There are three main modes of fermentation technique: batch, continuous, and fed-batch. A batch fermentation process is characterized by no addition to and withdrawal from the culture of biomass, fresh nutrient medium, and culture broth (with the exception of gas phase). In a continuous fermentation process, an open system is set up. Nutrient solution is added to the bioreactor continuously, and an equivalent amount of converted nutrient solution with microorganisms is simultaneously taken out of the system. In a fed-batch fermentation, substrate is added according to a predetermined feeding profile as the fermentation progresses.

A fed-batch operation may be followed by a terminal batch operation, with culture volume being equal to maximum permissible volume, to utilize the nutrients remaining in the culture at the end of fed-batch operation. A fed-batch operation is usually preceded by a batch operation. A typical run involving fed-batch operation therefore very often consists of the fed-batch operation sandwiched between two batch operations. This entire sequence (batch→fed-batch→batch) may be repeated many times leading to serial (or repeated) fed-batch operation.

Although fermentation operations are abundant and important in industries and academia which touch many human lives, high costs associated with many fermentation processes have become the bottleneck for further development and application of the products. Developing an economically and environmentally sound optimal cultivation method becomes the primary objective of fermentation process research nowadays. The goal is to control the process at its optimal state and to reach its maximum productivity with minimum development and production cost; in the meantime, the product quality should be maintained. A fermentation process may not be operated optimally for various reasons. For instance, an inappropriate nutrient feeding policy will result in a low production yield, even though the level of feeding rate is very high. An optimally controlled fermentation process offers the realization of high standards of product purity, operational safety, environmental regulations, and reduction in costs [246]. Nevertheless, different combinations and sequence of process conditions and medium components are needs to be biologically investigated to determine the growth condition that produces the biomass with the physiological state best constituted for product formation [195]. Moreover, the mathematical determination of optimal control in a fermentation process can be very difficult and open-ended due to frequent presence of nonlinearity in process models, inequality constraints on process variables, and implicit process discontinuities [17]. This presence gives rise to a multimodal and noncontinuous relation between a performance index and a control function.

Optimal control of fermentation processes has been a topic of research for many years. Considerable emphasis has been placed on the control of fed-batch fermenters because of their prevalence in industry [111, 129]. From a process operation point of view, most of studies are to calculate an optimal feed-rate profile that will optimize a given objective function. For the fed-batch process including one single operation, optimal control problem [34, 125, 231] and optimal adaptive control problem [18, 108, 247] have been discussed. Some useful tools such as Green function [193], the calculus of variations [135, 136, 179, 191], iterative dynamic programming [163], evolutionary algorithm [50, 210, 213], and genetic algorithm [217] are used to determine this profile in fed-batch processes. For the serial fed-batch operations, parameter optimization problem [252, 253, 278] and optimal impulsive control problem [84, 85, 254] have been reported. For the continuous process, time optimal control problem [60], maximum harvest problem [61, 62], optimal operation problem [239], and parameter optimization problem [226, 227] have been discussed. For the batch process, dynamic optimization problem [3, 245, 255, 256], optimal operation problem [34], and robust optimal control problem [183] have been investigated.

In this book, we focus on optimal control of fed-batch process including a serial of operations. This process is more complex and the abovementioned theories and methods are not applicable for this problem. Thus, new theory and computation methods are needed for the optimal control problems in this book.

### *1.3.2 1,3-Propanediol Fermentation*

Biodiesel (green diesel) fuels already constitute an alternative type of fuel for various types of diesel engines and heating systems [102]. Due to the increasing cost of conventional fuels, the application of biofuels in a large commercial scale is strongly recommended by various authorities, and this fact could likely result in the generation of tremendous quantities of glycerol in the near future [283]. Furthermore, besides biodiesel production units, concentrated glycerol-containing waters are also produced as the main by-product from fat saponification and alcoholic beverage fabrication units [16, 197]. For all of these reasons, glycerol overproduction and disposal is very likely to cause severe environmental problems in the near future. Therefore, conversion of glycerol to various higher-added-value products by the means of chemical and/or fermentation technology currently attracts much interest [31]. The most obvious target of biotechnological glycerol valorization is referred to its biotransformation into 1,3-propanediol (1,3-PD). This product is a substance of importance for the textile industry, due to its application as monomer for the synthesis of aliphatic polyesters [131]. Plastics based on this monomer exhibit good product properties [264]. Additionally, a recent development of a new polyester (polypropylene terephalate), presenting unique properties for the fiber industry, necessitated the drastic increase in the production of 1,3-PD [131]. Moreover, 1,3-PD can present various interesting applications in the chemical industry [31, 283].

1,3-PD is one of the oldest known fermentation products. It was reliably identified as early as in 1881 [83], in a glycerol fermentation mixed culture containing *Clostridium pasteurianum* as an active organism. The majority of commercial syntheses of 1,3-PD are from acrolein by Degussa (now owned by DuPont) and from ethylene oxide by Shell. Problems in these conventional processes are the high pressure applied in the hydroformylation and hydrogenation steps along with high temperature, use of expensive catalyst, and release of toxic intermediates. Considering the yield, product recovery, and environmental protection, much attention has been paid to its microbial production [49, 65, 185, 276]. The principal way of the biotechnological conversion of raw materials to 1,3-PD is referred to transformation of glycerol into 1,3-PD conducted by a number of microorganisms. The most extensively studied microorganisms belong to the species *Citrobacter freundii*, *Klebsiella pneumoniae* (*K. pneumoniae*), *Klebsiella oxytoca*, *Enterobacter agglomerans*, *Clostridium butyricum* and *Clostridium acetobutylicum* [196]. Among these organisms, *K. pneumoniae* is considered as one of the best "natural producers" and is paid more attention because of its appreciable substrate tolerance, yield, and productivity [173]. The enzymes and pathways involved in glycerol dissimilation to 1,3-PD production by *K. pneumoniae* have been elucidated in [82]. Regarding the fermentation, batch fermentation [16], continuous fermentation [173], and fed-batch fermentation [48, 287] have been performed. Substrate and product inhibitions are the main limiting factors for the microbial production of 1,3-PD by *K. pneumoniae*. In order to investigate the possibility of maximization of 1,3-PD production, genetically modified strains of the wild strain *K. pneumoniae* have been created [184, 286].

## *1.3.3 Kinetics and Physiological Modeling*

The optimization and control of bioprocesses often requires the establishment of a mathematical model that describes the metabolic activities of microorganisms, especially with respect to the responses of cells to a change in the physiological environment. Rate equations for microbial growth, substrate uptake, and product formation that describe the kinetics of a process are the basis for mathematical modeling. The rate equations used for microbial growth can be generally classified into two categories, i.e., unstructured models and structured models. The former treat a culture as a lumped quantity of biomass and does not consider intracellular components; the latter consider the heterogeneity of a culture and the intracellular components [13]. Despite impressive progress made recently in developing structured models for microbial growth [188], the unstructured models or semimechanistic models are still the most popular ones used in practice. The unstructured models include the most fundamental observations concerning microbial growth and are simple and easy to use, particularly for process control purposes.

The fermentation of glycerol by *K. pneumoniae* is a complex bioprocess, since microbial growth is subjected to multiple inhibitions of substrate and products, e.g., glycerol, 1,3-PD, ethanol, and acetate. The following kinetic model was proposed to describe microbial growth inhibited by several inhibitors [285]:

$$\mu = \mu_{\max} \frac{C_S}{K_S + C_S} \prod \left(1 - \frac{C_{P_i}}{C_{P_i}^*}\right)^{n_i} \tag{1.3}$$

where $\mu$ is the specific growth rate; $\mu_{\max}$ is the maximum specific growth rate; $C_S$ is the substrate concentration; $K_S$ is the saturation constant; $C_{P_i}$ is the concentration of inhibitor $P_i$; $C_{P_i}^*$ is the critical concentration of an inhibitor above which cells cease to grow; and $n_i$ is a constant. An excessive kinetics model was proposed in [282, 284]. In the excessive kinetics model, the specific substrate consumption rate ($q_S$) and the specific product formation rates ($q_{P_i}$, $P_i$ =1,3-PD, HAc, EtOH) of a substrate-sufficient culture could be expressed as follows:

$$q_S = m_S + \frac{\mu}{Y_S^m} + \Delta q_S^m \frac{C_S}{C_S + K_S^*}, \tag{1.4}$$

$$q_{P_i} = m_{P_i} + \mu Y_{P_i}^m + \Delta q_{P_i}^m \frac{C_S}{C_S + K_{P_i}^*},$$

$$(P_i = \text{1,3-PD, HAc, EtOH}) \tag{1.5}$$

where $Y_S^m$ and $m_S$ are the maximum growth yield and maintenance requirement of substrate under substrate-limited conditions, respectively; $\Delta q_S^m$ is the maximum increment of substrate consumption rate under substrate-sufficient conditions; $K_S^*$ is a saturation constant; $m_{P_i}$ and $Y_{P_i}^m$ are formation rate constants; $\Delta q_{P_i}^m$ is the maximum increase or decrease of product formation rate due to substrate excess; and $K_{P_i}^*$ is a saturation constant. An improved model was proposed to describe substrate consumption and product formation in a large range of feed glycerol concentrations in medium [271]. The main improvement is using the following expression to formulate the specific formation rate of ethanol $q_{\text{EtOH}}$:

$$q_{\text{EtOH}} = q_S \cdot \left( \frac{c_1}{c_2 + \mu C_S} + \frac{c_3}{c_4 + \mu C_S} \right) \tag{1.6}$$

where $c_1, c_2, c_3$, and $c_4$ are constants for determination of yield of ethanol on glycerol. Recently, the mathematical model describing the concentration changes of both extracellular substances and intracellular substances was proposed in [237].

## 1.4 Outline of the Book

The book is organized in eleven chapters. Except for Chap. 1 that briefly introduces the switched system, optimal control and fermentation process, and their literature reviews. Besides this short introduction, there are ten major chapters, which are briefly summarized as follows.

For the convenience of the reader, some mathematical preliminaries about measure theory and functional analysis are stated without proofs in Chap. 2. Engineers and applied scientists should be able to follow the mathematical proofs in the subsequent chapters with the aid of Chap. 2.

In Chap. 3, we review some results in constrained mathematical programming. This is important because after control parameterization, an optimal control problem is reduced to an optimal parameter selection problem, which is essentially a mathematical programming problem. Chapter 4 presents a crash course in optimal control theory for those readers who are not familiar with the subject.

From Chap. 5 onward, we focus our attention on the optimal control of switched systems arising in fermentation processes. We start from optimal control of a nonlinear multistage system, which is a degenerate switched system since switching law is decided a priori, in fed-batch fermentation process in Chap. 5. Compared with existing systems, the proposed system is much closer to the actual fermentation process. The optimal control model involving the nonlinear multistage system and subject to continuous state inequality constraint has been developed. The existence of optimal control is established by the theory of bounded variation. A global optimization algorithm based on the control parameterization concept and the improved particle swarm optimization algorithm is constructed to solve the optimal control problem. Numerical results show that the concentration of target product concentration at the terminal time is increased considerably compared with the experimental results.

In Chap. 6, we propose a switched autonomous system with variable switching instants to model the constantly fed-batch process. Taking the switching instants as the control function, we formulate an optimal control problem to optimize the fermentation process. By introducing a time-scaling transform, the optimal control problem is transcribed into an equivalent one with parameters and fixed switching instants. A computational approach to seek the optimal switching instants is developed. This method is based on the constraint transcription technique and the smoothing approximation method.

In Chap. 7, a time-dependent switched system, in which the feeding rate is the control function and the switching instants are the optimization variables, is proposed to formulate the fed-batch fermentation process. We then present a constrained optimal control problem involving the time-dependent switched system. To seek the optimal control and the optimal switching instants, we use the control parameterization enhancing transform together with the constraint transcription technique to convert the constrained optimal control problem into a sequence of mathematical programming problems. An improved particle swarm optimization is subsequently constructed to solve the resultant mathematical programming problem. Numerical results show that the target product concentration at the terminal time can be increased compared with previous results.

In Chap. 8, considering the hybrid nature in fed-batch fermentation process, we propose a state-based switched system to model the fermentation process. A constrained optimal switching control model is then presented. Because the number of the switchings is not known a priori, we reformulate the above optimal control problem as a two-level optimization problem. An optimization algorithm is developed to seek the optimal solution on the basis of a heuristic approach and the control parameterization method.

In Chap. 9, considering the microbial metabolism mechanism, i.e., the production of new biomass is delayed by the amount of time it takes to metabolize the nutrients, in fed-batch fermentation process, we propose a multistage time-delay system to formulate the process. In view of the effect of time delay and the high number of kinetic parameters in the system, the parametric sensitivity analysis is used to determine the key parameters. An optimal parameter selection model is presented, and a global optimization method is developed to seek the optimal key parameters. Numerical results show that the multistage time-delay system can describe the fed-batch fermentation process reasonably.

In Chap. 10, taking the mass of target product per unit time as the performance index, we formulate a constrained optimal control model with free terminal time to optimize the production process. Using a time-scale transformation, the optimal control problem is equivalently transcribed into the one with fixed terminal time. A computational approach is then developed to seek the optimal control and the optimal terminal time. This method is based on the control parameterization in conjunction with an improved differential evolution algorithm. Numerical results show that the mass of target product per unit time is increased considerably and the duration of the fermentation is shorted greatly compared with previous results.

In Chap. 11, taking the switching instants and the terminal time as the control variables, a free terminal time-delayed optimal control problem is proposed. Using a time-scaling transformation and parameterizing the switching instants into new parameters, an equivalently optimal control problem is presented. A numerical solution method is developed to seek the optimal control strategy by the smoothing approximation method and the gradient of the cost functional together with that of the constraints. Numerical results show that the mass of target product per unit time at the terminal time is increased considerably.

# Chapter 2
# Mathematical Preliminaries

For the convenience of the reader, some basic results in measure theory and functional analysis are presented without proofs in this chapter. The reader can turn to [57, 81, 99, 215, 260] for proofs of those theorems and for more detailed information.

## 2.1 Lebesgue Measure and Integration

For compactness of notation, we will refer to rectangular parallelepipeds in $\mathbb{R}^n$ whose sides are parallel to the coordinate axes simply as "boxes."

**Definition 2.1.** (a) A *box* in $\mathbb{R}^n$ is a set of the form

$$Q = [a_1, b_1] \times \cdots \times [a_n, b_n] = \prod_{i=1}^{n} [a_i, b_i]. \tag{2.1}$$

The *volume* of this box is

$$\mathrm{vol}(Q) = (b_1 - a_1) \cdots (b_n - a_n) = \prod_{i=1}^{n} (b_i - a_i). \tag{2.2}$$

(b) The *exterior Lebesgue measure* or simply *exterior measure* of a set $E \subseteq \mathbb{R}^n$ is

$$\mu^*(E) = \inf \left\{ \sum_k \mathrm{vol}(Q_k) \right\}, \tag{2.3}$$

where the infimum is taken over all finite or countable collections of boxes $Q_k$ such that $E \subseteq \bigcup_k Q_k$.

 
C. Liu, Z. Gong, *Optimal Control of Switched Systems Arising in Fermentation Processes*, Springer Optimization and Its Applications 97,
DOI 10.1007/978-3-662-43793-3_2

Thus, every subset of $\mathbb{R}^n$ has a uniquely defined exterior measure that lies in the range $0 \leqslant \mu^*(E) \leqslant +\infty$. Here are some of the basic properties of exterior measure.

*Property 2.1.* (a) If $Q$ is a box in $\mathbb{R}^n$, then $\mu^*(Q) = \mathrm{vol}(Q)$.
(b) If $E \subseteq F \subseteq \mathbb{R}^n$, then $\mu^*(E) \leqslant \mu^*(F)$.
(c) If $E_k \subseteq \mathbb{R}^n$ for $k \in \mathbb{N}$, then

$$\mu^*\left(\bigcup_{k=1}^{\infty} E_k\right) \leqslant \sum_{k=1}^{\infty} \mu^*(E_k). \tag{2.4}$$

(d) If $E \subseteq \mathbb{R}^n$ and $\boldsymbol{h} \in \mathbb{R}^n$, then $\mu^*(E+\boldsymbol{h}) = \mu^*(E)$, where $E+\boldsymbol{h} := \{\boldsymbol{x}+\boldsymbol{h} \mid \boldsymbol{x} \in E\}$.
(e) If $E \subseteq \mathbb{R}^n$ and $\epsilon > 0$, then there exists an open set $U \supseteq E$ such that $\mu^*(U) \leqslant \mu^*(E) + \epsilon$, and hence

$$\mu^*(E) = \inf\{\mu^*(U) \mid U \text{ is open and } U \supseteq E\}. \tag{2.5}$$

**Definition 2.2.** A set $E \subseteq \mathbb{R}^n$ is *Lebesgue measurable*, or simply *measurable*, if

$$\forall \epsilon > 0, \exists \text{ open } U \supseteq E \text{ such that } \mu^*(U - E) \leqslant \epsilon.$$

If $E$ is Lebesgue measurable, then its *Lebesgue measure* is its exterior Lebesgue measure and is denoted by $\mu(E) = \mu^*(E)$.

The following result summarizes some of the properties of Lebesgue measurable.

*Property 2.2.* Let $E$ and $E_k$ be measurable subsets of $\mathbb{R}^n$.

(a) If $E_1, E_2, \ldots$ are disjoint measurable subsets of $\mathbb{R}^n$, then

$$\mu\left(\bigcup_{k=1}^{\infty} E_k\right) = \sum_{k=1}^{\infty} \mu(E_k). \tag{2.6}$$

(b) If $E_1 \subseteq E_2$ and $\mu(E_1) < +\infty$, then $\mu(E_2 - E_1) = \mu(E_2) - \mu(E_1)$.
(c) If $E_1 \subseteq E_2 \subseteq \cdots$, then $\mu(\bigcup E_k) = \lim_{k\to\infty} \mu(E_k)$.
(d) If $E_1 \supseteq E_2 \supseteq \cdots$ and $\mu(E_1) < +\infty$, then $\mu(\bigcap E_k) = \lim_{k\to\infty} \mu(E_k)$.
(e) If $\boldsymbol{h} \in \mathbb{R}^n$, then $\mu(E + \boldsymbol{h}) = \mu(E)$, where $E + \boldsymbol{h} := \{\boldsymbol{x} + \boldsymbol{h} \mid \boldsymbol{x} \in E\}$.
(f) If $E \subseteq \mathbb{R}^m$ and $F \subseteq \mathbb{R}^n$ are measurable, then $E \times F \subseteq \mathbb{R}^{m+n}$ is measurable and $\mu(E \times F) = \mu(E)\mu(F)$.

The following concept is often used in the sequel.

**Definition 2.3.** A property that holds except possibly on a set of measure zero is said to hold *almost everywhere*, abbreviated a.e.

The essential supremum of a function is an example of a quantity that is defined in terms of a property that holds almost everywhere.

**Definition 2.4.** The *essential supremum* of a function $f : E \to \mathbb{R}$ is

$$\operatorname*{ess\,sup}_{x \in E} f(\boldsymbol{x}) = \inf\{M \mid f \leq M \text{ a.e.}\}. \tag{2.7}$$

We say that $f$ is *essentially bounded* if $\operatorname*{ess\,sup}_{x \in E} |f(\boldsymbol{x})| < \infty$.

Now, we define the class of measurable functions on subsets of $\mathbb{R}^n$.

**Definition 2.5.** Fix a measurable set $E \subseteq \mathbb{R}^n$, and let $f : E \to \mathbb{R}$ be given. Then $f$ is a *Lebesgue measurable function*, or simply a *measurable function*, if $f^{-1}(c, \infty) := \{\boldsymbol{x} \in E \mid f(\boldsymbol{x}) > c\}$ is a measurable subset of $\mathbb{R}^n$ for each $c \in \mathbb{R}$.

In particular, every continuous function $f : \mathbb{R}^n \to \mathbb{R}$ is measurable. However, a measurable function need not be continuous.

Measurability is preserved under most of the usual operations, including addition, multiplication, and limits.

*Property 2.3.* Let $E \subseteq \mathbb{R}^n$ be measurable.

(a) If $f : E \to \mathbb{R}$ is measurable and $g = f$ a.e., then $g$ is measurable.
(b) If $f, g : E \to \mathbb{R}$ are measurable, then so are $f + \alpha g (\alpha \in \mathbb{R})$, $f \cdot g$, $f/g$ $(g(\boldsymbol{x}) \neq 0)$, $\min\{f, g\}$, $\max\{f, g\}$, and $|f|$.
(c) If $f_n : E \to \mathbb{R}$ are measurable for $n \in \mathbb{N}$, then so are $\inf_n f_n$, $\sup_n f_n$, $\liminf_{n\to\infty} f_n$, and $\limsup_{n\to\infty} f_n$.

The following theorem says that pointwise convergence of measurable functions is uniform convergence on "most" of the set.

**Theorem 2.1 (Egoroff).** *Let $E \subseteq \mathbb{R}^n$ be measurable with $\mu(E) < \infty$. If $f_n, f : E \to \mathbb{R}$ are measurable functions and $f_n(\boldsymbol{x}) \to f(\boldsymbol{x})$ for a.e. $\boldsymbol{x} \in E$, then, for every $\epsilon > 0$, there exists a measurable set $E_\epsilon \subseteq E$ such that $\mu(E_\epsilon) < \epsilon$ and $f_n$ converges uniformly to $f$ on $E - E_\epsilon$, i.e.,*

$$\lim_{n\to\infty} \left( \sup_{x \notin E_\epsilon} | f_n(\boldsymbol{x}) - f(\boldsymbol{x})| \right) = 0. \tag{2.8}$$

To define the Lebesgue integral of a measurable function, we first begin with "simple functions."

**Definition 2.6.** Let $E \subseteq \mathbb{R}^n$ be measurable. A *simple function* on $E$ is a function $\varphi : E \to \mathbb{R}$ of the form

$$\varphi(\boldsymbol{x}) = \sum_{k=1}^{N} a_k \chi_{E_k}(\boldsymbol{x}), \tag{2.9}$$

where $N > 0$, $a_k \in \mathbb{R}$, $E_k$ is a measurable subset of $E$ and $\chi_{E_k} : E \to \mathbb{R}$ is the *indicator function* on $E_k$ defined by

$$\chi_{E_k}(\boldsymbol{x}) = \begin{cases} 1, & \text{if } \boldsymbol{x} \in E_k, \\ 0, & \text{otherwise.} \end{cases} \tag{2.10}$$

If $a_1, \dots, a_N \in \mathbb{R}$ are the distinct values assumed by a simple function $\varphi$ and we set $E_k = \{\boldsymbol{x} \in E \mid \varphi(\boldsymbol{x}) = a_k\}$, then $\varphi$ has the form given in Eq. (2.9) and the sets $E_1, \dots, E_N$ form a partition of $E$. We call this the *standard representation* of $\varphi$.

**Definition 2.7.** If $\varphi$ is a nonnegative simple function on $E$ with standard representation, then the *Lebesgue integral* of $\varphi$ over $E$ is

$$\int_E \varphi(\boldsymbol{x})\mathrm{d}\boldsymbol{x} = \sum_{k=1}^{N} a_k \mu(E_k). \tag{2.11}$$

**Definition 2.8.** If $f : E \to [0, \infty)$ is a measurable function, then the *Lebesgue integral* of $f$ over $E$ is

$$\int_E f(\boldsymbol{x})\mathrm{d}\boldsymbol{x} = \sup\left\{\int_E \varphi(\boldsymbol{x})\mathrm{d}\boldsymbol{x} \mid 0 \leqslant \varphi \leqslant f, \text{ and } \varphi \text{ is simple}\right\}. \tag{2.12}$$

Definition 2.8 is often cumbersome to implement. One application of the next result is that the integral of $f$ can be obtained as a limit instead of a supremum of integrals of simple functions. We say that a sequence of functions $\{f_n\}$ is *monotone increasing* if

$$f_1(\boldsymbol{x}) \leqslant f_2(\boldsymbol{x}) \leqslant \cdots, \text{ for all } \boldsymbol{x}. \tag{2.13}$$

**Theorem 2.2 (Lévi Monotone Convergence Theorem).** *Let $E \subseteq \mathbb{R}^n$ be measurable, and assume $\{f_n\}$ are nonnegative monotone increasing measurable functions on $E$ such that $f_n(\boldsymbol{x}) \to f(\boldsymbol{x})$ pointwise. Then*

$$\lim_{n\to\infty} \int_E f_n(\boldsymbol{x})\mathrm{d}\boldsymbol{x} = \int_E f(\boldsymbol{x})\mathrm{d}\boldsymbol{x}. \tag{2.14}$$

If we have functions $\{f_n\}$ that are not monotone increasing, then we may not be able to interchange a limit with an integral. The following result states that as long as $\{f_n\}$ are all nonnegative, we do at least have an inequality.

**Theorem 2.3 (Fatou's Lemma).** *If $\{f_n\}$ is a sequence of measurable, nonnegative functions on a measurable set $E \subseteq \mathbb{R}^n$, then*

$$\int_E \left(\liminf_{n\to\infty} f_n(\boldsymbol{x})\right) \mathrm{d}\boldsymbol{x} \leqslant \liminf_{n\to\infty} \int_E f_n(\boldsymbol{x})\mathrm{d}\boldsymbol{x}. \tag{2.15}$$

The following dominated convergence theorem is one of the most important convergence theorems for integrals.

**Theorem 2.4 (Lebesgue Dominated Convergence Theorem).** *Assume* $\{f_n\}$ *is a sequence of Lebesgue measurable functions on a measurable set* $E \subseteq \mathbb{R}^n$ *such that*

(a) $f(\boldsymbol{x}) = \lim_{n\to\infty} f_n(\boldsymbol{x})$ *exists for a.e.* $\boldsymbol{x} \in E$, *and*
(b) *there exists an integrable function* $g : E \to \mathbb{R}$ *such that*

$$|f_n(\boldsymbol{x})| \leqslant g(\boldsymbol{x}), \qquad \text{a.e. for every } n. \tag{2.16}$$

*Then* $f$ *is also integrable, and*

$$\lim_{n\to\infty} \int_E f_n(\boldsymbol{x})\mathrm{d}\boldsymbol{x} = \int_E f(\boldsymbol{x})\mathrm{d}\boldsymbol{x}. \tag{2.17}$$

## 2.2 Normed Spaces

We assume that the reader is familiar with vector spaces (which are also called *linear spaces*). The scalar field associated with the vector spaces will always be the real line $\mathbb{R}$. A norm on a vector space quantifies the idea of the "size" of a vector.

**Definition 2.9.** A vector space $X$ is called a *normed linear space* if for each $x \in X$, there is a (finite) real number $\|x\|$, called the *norm* of $x$, such that

(a) $\|x\| \geqslant 0$, for all $x \in X$, and $\|x\| = 0$ if and only if $x = 0$,
(b) $\|cx\| = |c| \cdot \|x\|$, for all $x \in X$ and scalar $c \in \mathbb{R}$, and
(c) $\|x + y\| \leqslant \|x\| + \|y\|$, for all $x, y \in X$.

Given a normed space $X$, it is usually clear from context what norm we mean to use on $X$. Therefore, we usually just write $\|\cdot\|$ to denote the norm on $X$. However, when there is a possibility of confusion, we may write $\|\cdot\|_X$ to specify that this norm is the norm on $X$.

**Definition 2.10.** Let $X$ be a normed linear space.

(a) A sequence of vectors $\{x_n\}$ in $X$ *converges* to $x \in X$ if $\lim_{n\to\infty} \|x_n - x\| = 0$, i.e., if

$$\forall \epsilon > 0,\ \exists N > 0,\ \forall n \geqslant N,\ \|x_n - x\| < \epsilon. \tag{2.18}$$

In this case, we write either $x_n \to x$ or $\lim_{n\to\infty} x_n = x$.

(b) A sequence of vectors $\{x_n\}$ in $X$ is a *Cauchy sequence* in $X$ if $\lim_{m,n\to\infty} \|x_m - x_n\| = 0$. More precisely, this means that

$$\forall \epsilon > 0,\ \exists N > 0,\ \forall m, n \geqslant N,\ \|x_m - x_n\| < \epsilon. \tag{2.19}$$

Every convergent sequence in a normed space is a Cauchy sequence. However, the converse is not true in general.

**Definition 2.11.** A normed space $X$ is *complete* if it is the case that every Cauchy sequence in $X$ is a convergent sequence. A complete normed linear space is called a *Banach space*.

The simplest example of a Banach space is the scalar field $\mathbb{R}$, where the norm on $\mathbb{R}$ is the absolute value. The next example of a Banach space is $\mathbb{R}^n$, the set of all $n$-tuples of scalars, where $n$ is a positive integer. There are many choices of norms for $\mathbb{R}^n$. Writing a generic vector $\boldsymbol{v} \in \mathbb{R}^n$ as $\boldsymbol{v} := (v_1, v_2, \ldots, v_n)^\top$, each of the following defines a norm on $\mathbb{R}^n$, and $\mathbb{R}^n$ is complete with respect to each of these norms:

$$\|\boldsymbol{v}\|_p := \begin{cases} (|v_1|^p + |v_2|^p + \cdots + |v_n|^p)^{1/p}, & 1 \leqslant p < \infty, \\ \max\{|v_1|, |v_2|, \ldots, |v_n|\}, & p = \infty. \end{cases} \tag{2.20}$$

The *Euclidean norm* $\|\boldsymbol{v}\|$ of a vector $\boldsymbol{v} \in \mathbb{R}^n$ is the norm corresponding to the choice $p = 2$, i.e.,

$$\|\boldsymbol{v}\| = \|\boldsymbol{v}\|_2 := \sqrt{|v_1|^2 + |v_2|^2 + \cdots + |v_n|^2}.$$

In fact, there can be many norms on a given Banach space.

**Definition 2.12.** Suppose that $X$ is a normed linear space with respect to a norm $\|\cdot\|$ and also with respect to another norm $|||\cdot|||$. These norms are *equivalent* if there exist constants $C_1, C_2 > 0$ such that

$$\forall x \in X,\ C_1\|x\| \leqslant |||x||| \leqslant C_2\|x\|. \tag{2.21}$$

Note that if $\|\cdot\|$ and $|||\cdot|||$ are equivalent norms on $X$, then they define the same convergence criterion in the sense that

$$\lim_{n\to\infty} \|x_n - x\| = 0 \quad \Longleftrightarrow \quad \lim_{n\to\infty} |||x_n - x||| = 0.$$

Any two of the norms $\|\cdot\|_p$ on $\mathbb{R}^n$ are equivalent. This is a special case of the following theorem.

**Theorem 2.5.** *If $V$ is a finite-dimensional vector space, then any two norms on $V$ are equivalent.*

Now we give two examples of infinite-dimensional Banach spaces. One is $l^p$ space whose elements are infinite sequences of scalars, and another is $L_p$ space whose elements are measurable functions on $I \subseteq \mathbb{R}$:

(i) $l_p$ space

$$l_p := \begin{cases} \left\{ \{x_k\}_{k\in\mathbb{N}} \mid x_k \in \mathbb{R} \text{ and } \sum_k |x_k|^p < +\infty \right\}, & p \in [1, +\infty), \\ \left\{ \{x_k\}_{k\in\mathbb{N}} \mid x_k \in \mathbb{R} \text{ and } \sup_k |x_k|^p < +\infty \right\}, & p = +\infty, \end{cases} \tag{2.22}$$

with the norm

$$\|\{x_k\}\|_{l_p} := \begin{cases} \left( \sum_k |x_k|^p \right)^{1/p}, & p \in [1, +\infty), \\ \sup_k |x_k|, & p = +\infty. \end{cases} \tag{2.23}$$

(ii) $L_p$ space

$$L_p(I, \mathbb{R}^n) := \begin{cases} \left\{ f : I \to \mathbb{R}^n \mid \int_I \|f(t)\|^p \mathrm{d}t < \infty \right\}, & p \in [1, +\infty), \\ \left\{ f : I \to \mathbb{R}^n \mid \operatorname{ess\,sup}_{t\in I} \|f(t)\| < \infty \right\}, & p = +\infty, \end{cases} \tag{2.24}$$

with the norm

$$\|f\|_{L_p} := \begin{cases} \left( \int_I \|f(t)\|^p \mathrm{d}t \right)^{1/p}, & p \in [1, +\infty), \\ \operatorname{ess\,sup}_{t\in I} \|f(t)\|, & p = +\infty. \end{cases} \tag{2.25}$$

It is well-known that if $I$ is a finite interval, then

(a) $\|f\|_{L_\infty} = \lim_{p\to\infty} \left( \int_I \|f(t)\|^p \mathrm{d}t \right)^{1/p}$; and
(b) $L_1(I, \mathbb{R}^n) \supset L_2(I, \mathbb{R}^n) \supset \cdots \supset L_\infty(I, \mathbb{R}^n)$.

In optimal control theory, we shall be concerned with the space $C(I, \mathbb{R}^n)$ of all continuous functions from $I \subseteq \mathbb{R}$ to $\mathbb{R}^n$. The space $C(I, \mathbb{R}^n)$ is a vector space and becomes a Banach space when it is equipped with the *sup norm* defined by

$$\|f\|_{C(I,\mathbb{R}^n)} := \sup_{t\in I} \|f(t)\|, \tag{2.26}$$

where $f := (f_1, f_2, \ldots, f_n)^\top$ and $\|f(t)\| := \sqrt{f_1^2(t) + f_2^2(t) + \cdots + f_n^2(t)}$.

A set $A \subseteq C(I, \mathbb{R}^n)$ is said to be *equicontinuous* if for any $\epsilon > 0$, there exists a $\delta > 0$ such that for all $\boldsymbol{f} \in A$,

$$\left\| \boldsymbol{f}(t') - \boldsymbol{f}(t) \right\| < \epsilon \tag{2.27}$$

whenever $t', t \in I$ are such that $|t' - t| < \delta$.

Let $I := [a, b] \subset \mathbb{R}$ and $\boldsymbol{f} := (f_1, f_2, \ldots, f_n)^\top \in C(I, \mathbb{R}^n)$. The function $\boldsymbol{f}$ is said to be *absolutely continuous* on $I$ if for any given $\epsilon > 0$, there exists a $\delta > 0$ such that

$$\sum_{k=1}^{m} \left\| \boldsymbol{f}(t_k) - \boldsymbol{f}\left(t'_k\right) \right\| < \epsilon \tag{2.28}$$

for every finite collection $\{(t_k, t'_k)\}$ of non-overlapping intervals satisfying

$$\sum_{k=1}^{m} \left| t_k - t'_k \right| < \delta. \tag{2.29}$$

The class of all such absolutely continuous functions is denoted by $AC(I, \mathbb{R}^n)$. Clearly, a Lipschitz continuous function on $I$ is absolutely continuous.

**Theorem 2.6.** *If $\boldsymbol{f} \in L_1(I, \mathbb{R}^n)$ and $\boldsymbol{g}$ is defined by*

$$\boldsymbol{g}(t) = \boldsymbol{g}(a) + \int_a^t \boldsymbol{f}(\vartheta)\mathrm{d}\vartheta, \quad t \in I, \tag{2.30}$$

*then $\boldsymbol{g} \in AC(I, \mathbb{R}^n)$ and $\dfrac{\mathrm{d}\boldsymbol{g}(t)}{\mathrm{d}t} = \boldsymbol{f}(t)$ a.e. on $I$.*

## 2.3 Linear Functionals and Dual Spaces

**Definition 2.13.** Let $X$ be a normed linear space. A map $f : X \to \mathbb{R}$ is called a *bounded linear functional* if

$$f(\alpha x + \beta y) = \alpha f(x) + \beta f(y), \quad \forall \alpha, \beta \in \mathbb{R}, x, y \in X, \tag{2.31}$$

and there exists a constant $M > 0$ such that

$$|f(x)| \leqslant M \|x\|, \quad \forall x \in X. \tag{2.32}$$

A functional $f$ which only satisfies (2.31) is bounded if and only if it is continuous.

**Definition 2.14.** Given a normed linear space $X$, the space of all bounded linear functionals on $X$ with norm

$$\|f\| := \sup_{x \in X \setminus \{0\}} \frac{|f(x)|}{\|x\|}, \tag{2.33}$$

is the *dual space* of $X$ and is denoted by

$$X^* = \{f : X \to \mathbb{R} |\ f \text{ is a bounded linear functional}\} \tag{2.34}$$

Since $\mathbb{R}$ is complete, the dual space $X^*$ of a normed space $X$ is complete, even if $X$ is not. The dual of $X^*$, also known as the *second dual* of $X$, is denoted by $X^{**}$.

**Definition 2.15.** A Banach space $X$ is called *reflexive* if $X = X^{**}$.

Note that $X^{**}$ is a Banach space, so $X$ must be a Banach space if we are to be able to identify $X$ with $X^{**}$.

**Theorem 2.7.** *A Banach space $X$ is reflexive if and only if $X^*$ is reflexive.*

**Theorem 2.8.** *Suppose that the Banach space $X$ is not reflexive. Then the inclusions $X \subseteq X^{**} \subseteq X^{****} \subseteq \cdots$ and $X^* \subseteq X^{***} \subseteq \cdots$ are all strict.*

The following theorem gives the duals of some of the classical Banach spaces.

**Theorem 2.9.** *For any $p \in [1, +\infty)$,*

$$(l_p)^* = l_q, \tag{2.35}$$

$$\{L_p(I, \mathbb{R}^n)\}^* = L_q(I, \mathbb{R}^n), \tag{2.36}$$

*where*

$$q := \begin{cases} \dfrac{p}{p-1}, & \text{if } p \neq 1, \\ +\infty, & \text{if } p = 1. \end{cases} \tag{2.37}$$

*Furthermore, for each $p \in (1, +\infty)$, $l_p$ and $L_p(I, \mathbb{R}^n)$ are all reflexive.*

It should be noted that $(l_\infty)^* \neq l_1$ and $\{L_\infty(I, \mathbb{R}^n)\}^* \neq L_1(I, \mathbb{R}^n)$.

Finally, we give some types of convergence. Part (a) of the following definition recalls the usual notion of convergence as given in Definition 2.10, and parts (b) and (c) introduce some new types of convergence.

**Definition 2.16.** Let $X$ be a Banach space.

(a) A sequence $\{x_n\}$ of elements of $X$ *converges* (strongly) to $x \in X$ if

$$\lim_{n \to \infty} \|x_n - x\| = 0. \tag{2.38}$$

We denote this convergence by $\lim_{n \to \infty} x_n = x$.

(b) A sequence $\{x_n\}$ of elements of $X$ *converges weakly* to $x \in X$ if

$$\lim_{n\to\infty} f(x_n - x) = 0, \quad \forall f \in X^*. \tag{2.39}$$

We denote weak convergence by $w - \lim_{n\to\infty} x_n = x$.

(c) A sequence $\{f_n\}$ of functionals in $X^*$ *converges weak** to $f \in X^*$ if

$$\lim_{n\to\infty} f_n(x) = f(x), \quad \forall x \in X. \tag{2.40}$$

We denote weak* convergence by $w^* - \lim_{n\to\infty} f_n(x) = f(x)$.

Note that weak* convergence only applies to convergence of functionals in a dual space $X^*$. Every weakly convergent sequence is bounded. Strong convergence in $X$ implies weak convergence in $X$ and weak convergence in $X^*$ implies weak* convergence in $X^*$.

**Theorem 2.10 (Banach–Saks–Mazur).** *Let $X$ be a normed space and $\{x_n\}$ be a sequence in $X$ converging weakly to $x$. Then there exists a sequence of finite convex combinations of $\{x_n\}$ that converges strongly to $x$.*

## 2.4 Bounded Variation

By a partition of the interval $I := [a, b] \subset \mathbb{R}$, we mean a finite set of points $t_i \in I$, $i = 0, 1, \ldots, m$, such that

$$a = t_0 < t_1 < t_2 < \cdots < t_m = b. \tag{2.41}$$

A function $h$ defined on $I$ is said to be of *bounded variation* if there is a constant $K > 0$ such that for any partition of $I$,

$$\sum_{i=1}^{m} |h(t_i) - h(t_{i-1})| \leqslant K. \tag{2.42}$$

The *total variation* of $h$, denoted by $\bigvee_a^b h(t)$, is defined by

$$\bigvee_a^b h(t) = \sup \sum_{i=1}^{m} |h(t_i) - h(t_{i-1})|, \tag{2.43}$$

where the supremum is taken with respect to all partitions of $I$. The total variation of a constant function is zero and the total variation of a monotonic function is the absolute value of the difference between the function values at the endpoints $a$ and $b$.

The space $BV(I)$ is defined as the space of all functions of bounded variation on $I$ together with the norm defined by

$$\|h\|_{BV} = |h(a)| + \bigvee_a^b h(t). \tag{2.44}$$

Suppose $h \in BV(I)$. Then, $h$ is differentiable a.e. on $I$. If $h : I \to \mathbb{R}$ is absolutely continuous, then it is of bounded variation.

**Theorem 2.11.** *If $h \in BV(I)$, then $h$ is absolutely continuous if and only if*

$$\int_a^b \left|\dot{h}(t)\right| \mathrm{d}t = \bigvee_a^b h(t). \tag{2.45}$$

If $h$ is monotone, then $h \in BV(I)$ and $\bigvee_a^b h(t) = |h(b) - h(a)|$.

We now consider a function $\boldsymbol{h} := (h_1, h_2, \ldots, h_n)^\top : I \to \mathbb{R}^n$. The *full variation* of $\boldsymbol{h}$ is defined as

$$\bigvee_a^b \boldsymbol{h}(t) = \sum_{i=1}^n \bigvee_a^b h_i(t). \tag{2.46}$$

Let $BV(I, \mathbb{R}^n)$ be the space of all functions $\boldsymbol{h} : I \to \mathbb{R}^n$ which are of bounded variation on $I$.

**Theorem 2.12.** *If $\boldsymbol{h} \in BV(I, \mathbb{R}^n)$, then $\boldsymbol{h}(t+0) := \lim_{s \downarrow t} \boldsymbol{h}(s)$, the limit from the right at $t$, exists if $a \leqslant t < b$; and $\boldsymbol{h}(t-0) := \lim_{s \uparrow t} \boldsymbol{h}(s)$, the limit from the left at $t$, exists if $a < t \leqslant b$.*

In order that $\boldsymbol{h}$ approaches a limit in $\mathbb{R}^n$ as $s$ approaches $t$ from the right (respectively, from the left), the following condition is necessary and sufficient: For each $\epsilon > 0$, there corresponds a $\delta > 0$ such that $\|\boldsymbol{h}(\tau) - \boldsymbol{h}(t)\| < \epsilon$ if $s < \tau < t + \delta$ (respectively, $t - \delta < \tau < s$).

**Theorem 2.13.** *If $\boldsymbol{h} \in BV(I, \mathbb{R}^n)$, the set of points of discontinuity of $\boldsymbol{h}$ is countable.*

Let $E$ be a family of functions in $BV(I,\mathbb{R}^n)$. It is said to be *equibounded with equibounded total variation* if there exist constants $K_1 > 0, K_2 > 0$ such that $\|\boldsymbol{h}(t)\| \leqslant K_1$ and $\bigvee_a^b \boldsymbol{h}(t) \leqslant K_2$ for all $\boldsymbol{h} \in E$.

**Theorem 2.14 (Helly).** *Let $E$ be a family of functions in $BV(I,\mathbb{R}^n)$ which is equibounded with equibounded total variation. Then, any sequence $\{\boldsymbol{h}^i\}$ of elements in $E$ contains a subsequence $\{\boldsymbol{h}^{i_k}\}$ which converges pointwise everywhere on $I$ toward a function $\boldsymbol{h}^0 \in BV(I,\mathbb{R}^n)$ with*

$$\bigvee_a^b \boldsymbol{h}^0(t) \leqslant \liminf_{k\to\infty} \bigvee_a^b \boldsymbol{h}^{i_k}(t). \tag{2.47}$$

# Chapter 3
# Constrained Mathematical Programming

The optimal control problem can be reduced to optimal parameter selection problems by approximating the control functions with an appropriate series of spline functions. Although the constraint on the dynamical system still exists, the problem may, after the parameterization, be viewed as an implicit mathematical programming problem. The solution to the optimal control problem may thus be obtained through solving a sequence of resulting mathematical programming problems, although the computational procedure is much more involved. Thus, understanding of the fundamental concepts, theories, and methods of mathematical programming is obviously important.

To begin, we note that the notation used in this chapter is applicable only to this chapter. For example, subsequent appearance of $\boldsymbol{x}$ is not to be confused with the state vector in other chapters.

## 3.1 Introduction

As opposed to optimal control problems, mathematical programming problems are static in nature. The general constrained mathematical programming problem is described by

$$\min_{\boldsymbol{x} \in \mathbb{R}^n} \quad f(\boldsymbol{x}) \tag{3.1}$$

$$\text{s.t.} \quad c_i(\boldsymbol{x}) = 0, \quad i \in \mathcal{E}, \tag{3.2}$$

$$c_i(\boldsymbol{x}) \geqslant 0, \quad i \in \mathcal{I}, \tag{3.3}$$

where $f$ and the functions $c_i$ are all smooth, real-valued functions on a subset of $\mathbb{R}^n$ and $\mathcal{E}$ and $\mathcal{I}$ are two finite sets of indices. $f$ is the *objective function*, while $c_i$,



C. Liu, Z. Gong, *Optimal Control of Switched Systems Arising in Fermentation Processes*, Springer Optimization and Its Applications 97,
DOI 10.1007/978-3-662-43793-3_3

$i \in \mathcal{E}$ are the *equality constraints* and $c_i$, $i \in \mathcal{I}$ are the *inequality constraints*. We define the *feasible set* $\Omega$ to be the set of points $\boldsymbol{x}$ that satisfy the constraints, that is,

$$\Omega = \{\boldsymbol{x} \,|\, c_i(\boldsymbol{x}) = 0, i \in \mathcal{E};\ c_i(\boldsymbol{x}) \geqslant 0, i \in \mathcal{I}\}, \tag{3.4}$$

so that we can write (3.1)–(3.3) more compactly as

$$\min_{\boldsymbol{x} \in \Omega} \ f(\boldsymbol{x}). \tag{3.5}$$

We have the following definitions of the different types of solutions for (3.1)–(3.3).

**Definition 3.1.** A vector $\boldsymbol{x}^*$ is a *local solution* of the problem (3.1)–(3.3) if $\boldsymbol{x}^* \in \Omega$ and there is a neighborhood $\mathcal{N}$ of $\boldsymbol{x}^*$ such that $f(\boldsymbol{x}) \geqslant f(\boldsymbol{x}^*)$ for $\boldsymbol{x} \in \mathcal{N} \cap \Omega$.

**Definition 3.2.** A vector $\boldsymbol{x}^*$ is a *strict local solution* of the problem (3.1)–(3.3) if $\boldsymbol{x}^* \in \Omega$ and there is a neighborhood $\mathcal{N}$ of $\boldsymbol{x}^*$ such that $f(\boldsymbol{x}) > f(\boldsymbol{x}^*)$ for $\boldsymbol{x} \in \mathcal{N} \cap \Omega$ with $\boldsymbol{x} \neq \boldsymbol{x}^*$.

**Definition 3.3.** A vector $\boldsymbol{x}^*$ is a *global solution* of the problem (3.1)–(3.3) if $\boldsymbol{x}^* \in \Omega$ and $f(\boldsymbol{x}) \geqslant f(\boldsymbol{x}^*)$ for all $\boldsymbol{x} \in \Omega$.

The following definition is an important terminology in the constrained mathematical programming.

**Definition 3.4.** The *active set* $\mathcal{A}(\boldsymbol{x})$ at any feasible $\boldsymbol{x}$ consists of the equality constraint indices from $\mathcal{E}$ together with the indices of the inequality constraints $i$ for which $c_i(\boldsymbol{x}) = 0$, that is,

$$\mathcal{A}(\boldsymbol{x}) = \mathcal{E} \cup \{i \in \mathcal{I} |\, c_i(\boldsymbol{x}) = 0\}. \tag{3.6}$$

At a feasible point $\boldsymbol{x}$, the inequality constraint $i \in \mathcal{I}$ is said to be *active* if $c_i(\boldsymbol{x}) = 0$ and *inactive* if the strict inequality $c_i(\boldsymbol{x}) > 0$ is satisfied.

Optimization techniques, or algorithms, are used to find the solution to the problem specified in (3.1)–(3.3). Note that, for many problems, more than one optimum may exist. There are many options for classifying the available optimization techniques. A short overview of the available algorithms, using a broad classification as either gradient-based or evolutionary algorithms, is presented.

## 3.2 Gradient-Based Algorithms

We shall first summarize the main optimality conditions without proofs. Then, we introduce three gradient-based algorithms, i.e., the quadratic penalty method, augmented Lagrangian method, and sequential quadratic programming (SQP) method, for solving the constrained mathematical programming problem (3.1)–(3.3). For further details, the reader is referred to [19, 80, 162, 189].

### 3.2.1 Optimality Conditions

The principal tool used in analyzing the constrained mathematical programming problem is the Kuhn–Tucker theory. For completeness, we need the following preliminaries.

**Definition 3.5.** The point $\boldsymbol{x}$ is said to be a *regular point* of the constraints (3.2)–(3.3) if $\boldsymbol{x}^*$ satisfies all the constraints and if the gradients of the equality and active inequality constraints

$$\{\nabla c_i(\boldsymbol{x}^*), i \in \mathcal{A}(\boldsymbol{x}^*)\} \tag{3.7}$$

are linearly independent, where

$$\nabla c_i(\boldsymbol{x}^*) := \left.\frac{\partial c_i(\boldsymbol{x})}{\partial \boldsymbol{x}}\right|_{\boldsymbol{x}=\boldsymbol{x}^*}. \tag{3.8}$$

Note that condition (3.7) is known as a *constraint qualification*.

As a preliminary to stating the necessary conditions for $\boldsymbol{x}^*$ to be a local minimizer, we define the *Lagrangian function* for the problem (3.1)–(3.3)

$$L(\boldsymbol{x}, \boldsymbol{\lambda}) = f(\boldsymbol{x}) - \sum_{i \in \mathcal{E} \cup \mathcal{I}} \lambda_i c_i(\boldsymbol{x}). \tag{3.9}$$

Noting that $\nabla_x L(\boldsymbol{x}, \boldsymbol{\lambda}) = \nabla f(\boldsymbol{x}) - \sum_{i \in \mathcal{E} \cup \mathcal{I}} \lambda_i \nabla c_i(\boldsymbol{x})$.

The necessary conditions defined in the following theorem are called first-order conditions. These conditions are the foundation for many of the algorithms.

**Theorem 3.1.** *Suppose that $\boldsymbol{x}^*$ is a local solution of (3.1)–(3.3) and also a regular point for the constraints. Then there is a Lagrange multiplier vector $\boldsymbol{\lambda}^*$, with components $\lambda_i$, $i \in \mathcal{E} \cup \mathcal{I}$, such that the following conditions are satisfied at $(\boldsymbol{x}^*, \boldsymbol{\lambda}^*)$:*

$$\nabla_x L(\boldsymbol{x}^*, \boldsymbol{\lambda}^*) = \mathbf{0}, \tag{3.10a}$$

$$c_i(\boldsymbol{x}^*) = 0, \quad \textit{for all } i \in \mathcal{E}, \tag{3.10b}$$

$$c_i(\boldsymbol{x}^*) \geq 0, \quad \textit{for all } i \in \mathcal{I}, \tag{3.10c}$$

$$\lambda_i^* c_i(\boldsymbol{x}^*) = 0, \ \lambda_i^* \geq 0, \quad \textit{for all } i \in \mathcal{I}. \tag{3.10d}$$

The conditions (3.10) are often known as the Karush–Kuhn–Tucker *conditions*, or KKT *conditions* for short. Note that condition (3.10d) implies that if the $i$th inequality constraint is inactive, then $\lambda_i = 0$, and conversely, if $\lambda_i > 0$, then the $i$th inequality constraint must be active.

We turn now to the linearized feasible direction set, which we define as follows.

**Definition 3.6.** Given a feasible point $\boldsymbol{x}$ and the active constraint set $\mathcal{A}(\boldsymbol{x})$, the set of linearized feasible directions $\mathcal{F}(\boldsymbol{x})$ is

$$\mathcal{F}(\boldsymbol{x}) = \left\{\boldsymbol{d} \mid \boldsymbol{d}^{\top}\nabla c_i(\boldsymbol{x}) = 0, \text{ for all } i \in \mathcal{E}, \boldsymbol{d}^{\top}\nabla c_i(\boldsymbol{x}) \geqslant 0, \text{ for all } i \in \mathcal{A}(\boldsymbol{x}) \cap \mathcal{I}\right\}. \tag{3.11}$$

The next theorem gives a necessary condition involving the second derivatives.

**Theorem 3.2.** *Suppose that* $\boldsymbol{x}^*$ *is a local solution of* (3.1)–(3.3) *and also a regular point for the constraints. Let* $\boldsymbol{\lambda}^*$ *be the Lagrange multiplier vector for which the* KKT *conditions* (3.10) *are satisfied. Then*

$$\boldsymbol{d}^{\top}\nabla^2_{xx}L\left(\boldsymbol{x}^*, \boldsymbol{\lambda}^*\right)\boldsymbol{d} \geqslant 0, \quad \text{for all } \boldsymbol{d} \in \mathcal{C}\left(\boldsymbol{x}^*, \boldsymbol{\lambda}^*\right), \tag{3.12}$$

*where*

$$\mathcal{C}\left(\boldsymbol{x}^*, \boldsymbol{\lambda}^*\right) := \left\{\boldsymbol{d} \in \mathcal{F}\left(\boldsymbol{x}^*\right) | \nabla c_i\left(\boldsymbol{x}^*\right)^{\top}\boldsymbol{d} = 0, \text{for all } i \in \mathcal{A}\left(\boldsymbol{x}^*\right) \cap \mathcal{I} \text{ with } \lambda_i^* > 0\right\}. \tag{3.13}$$

The second-order sufficient condition stated in the next theorem looks very much like the necessary condition just discussed, but it differs in that the constraint qualification is not required and the inequality in (3.12) is replaced by a strict inequality.

**Theorem 3.3.** *Suppose that for some feasible point* $\boldsymbol{x}^* \in \mathbb{R}^n$, *there is a Lagrange multiplier vector* $\boldsymbol{\lambda}^*$ *such that the* KKT *conditions* (3.10) *are satisfied. Suppose also that*

$$\boldsymbol{d}^{\top}\nabla^2_{xx}L\left(\boldsymbol{x}^*, \boldsymbol{\lambda}^*\right)\boldsymbol{d} > 0, \quad \text{for all } \boldsymbol{d} \in \mathcal{C}\left(\boldsymbol{x}^*, \boldsymbol{\lambda}^*\right), \ \boldsymbol{d} \neq \boldsymbol{0}. \tag{3.14}$$

*Then* $\boldsymbol{x}^*$ *is a strict local solution for* (3.1)–(3.3).

### 3.2.2 *The Quadratic Penalty Method*

The penalty methods for constrained mathematical programming replace the original problem by a sequence of subproblems in which the constraints are represented by terms added to the objective. The simplest penalty function of this type is the *quadratic penalty function*, in which the penalty terms are the squares of the constraint violations.

We describe this approach first in the context of the equality-constrained problem

$$\min_{x \in \mathbb{R}^n} \; f(\boldsymbol{x}) \tag{3.15a}$$

$$\text{s.t.} \;\; c_i(\boldsymbol{x}) = 0, \quad i \in \mathcal{E}, \tag{3.15b}$$

which is a special case of (3.1)–(3.3). The quadratic penalty function $Q(\boldsymbol{x};\mu)$ for this formulation is

$$Q(\boldsymbol{x};\mu) = f(\boldsymbol{x}) + \frac{\mu}{2}\sum_{i \in \mathcal{E}} c_i^2(\boldsymbol{x}), \tag{3.16}$$

where $\mu > 0$ is the *penalty parameter*. By driving $\mu$ to $\infty$, we penalize the constraint violations with increasing severity. It makes good intuitive sense to consider a sequence of values $\{\mu_k\}$ with $\mu_k \to \infty$ as $k \to \infty$ and to seek the approximate minimizer $\boldsymbol{x}_k$ of $Q(\boldsymbol{x};\mu_k)$ for each $k$.

For the general constrained mathematical programming problem (3.1)–(3.3), we can define the quadratic penalty function as

$$Q(\boldsymbol{x};\mu) = f(\boldsymbol{x}) + \frac{\mu}{2}\sum_{i \in \mathcal{E}} c_i^2(\boldsymbol{x}) + \frac{\mu}{2}\sum_{i \in \mathcal{I}} \left([c_i(\boldsymbol{x})]^-\right)^2, \tag{3.17}$$

where $[y]^-$ denotes $\max\{-y, 0\}$. In this case, $Q$ may be less smooth than the objective and constraint functions. For instance, if one of the inequality constraints is $x_1 \geqslant 0$, then the function $\min^2\{0, x_1\}$ has a discontinuous second derivative, so that $Q$ is no longer twice continuously differentiable.

We describe some convergence properties of the quadratic penalty method in the following two theorems. We restrict our attention to the equality-constrained problem (3.15), for which the quadratic penalty function is defined by (3.16).

For the first result we assume that the penalty function $Q(\boldsymbol{x};\mu_k)$ has a (finite) minimizer for each value of $\mu_k$.

**Theorem 3.4.** *Suppose that each $\boldsymbol{x}_k$ is the exact global minimizer of $Q(\boldsymbol{x};\mu_k)$ defined by* (3.16) *and that $\mu_k \to \infty$. Then every limit point $\boldsymbol{x}^*$ of the sequence $\{\boldsymbol{x}_k\}$ is a global solution of the problem* (3.15).

Since this result requires us to find the global minimizer for each subproblem, this desirable property of convergence to the global solution of (3.15) cannot be attained in general. The next result concerns convergence properties of the sequence $\{\boldsymbol{x}_k\}$ when we allow inexact (but increasingly accurate) minimizations of $Q(\cdot;\mu_k)$. We make the assumption that the stop test $\|\nabla_x Q(\boldsymbol{x};\mu_k)\| \leqslant \tau_k$ is satisfied for all $k$.

**Theorem 3.5.** *Suppose that the tolerances and penalty parameters satisfy $\tau_k \to 0$ and $\mu_k \to \infty$. Then if a limit point $\boldsymbol{x}^*$ of the sequence $\{\boldsymbol{x}_k\}$ is infeasible, it is a stationary point of the function $\|\boldsymbol{c}(\boldsymbol{x})\|^2$. On the other hand, if a limit point $\boldsymbol{x}^*$ is feasible and also a regular point for the constraints, then $\boldsymbol{x}^*$ is a* KKT *point for the* *problem* (3.15). *For such points, we have for any infinite subsequence* $\mathcal{K}$ *such that* $\lim_{k\in\mathcal{K}} \boldsymbol{x}_k = \boldsymbol{x}^*$ *that*

$$\lim_{k\in\mathcal{K}} -\mu_k c_i(\boldsymbol{x}_k) = \lambda_i^*, \quad \textit{for all } i \in \mathcal{E}, \tag{3.18}$$

*where* $\boldsymbol{\lambda}^*$ *is the multiplier vector that satisfies the* KKT *conditions* (3.10) *for the equality-constrained problem* (3.15).

### 3.2.3 Augmented Lagrangian Method

The *augmented Lagrangian method* or *multipliers penalty method* by introducing explicit Lagrange multiplier estimates into the objective can preserve the smoothness and reduce the possibility of ill conditioning.

We describe this approach first in the context of the equality-constrained problem (3.15). The augmented Lagrangian function $L_A(\boldsymbol{x}, \boldsymbol{\lambda}; \mu)$ for this formulation is

$$L_A(\boldsymbol{x}, \boldsymbol{\lambda}; \mu) = f(\boldsymbol{x}) - \sum_{i\in\mathcal{E}} \lambda_i c_i(\boldsymbol{x}) + \frac{\mu}{2} \sum_{i\in\mathcal{E}} c_i^2(\boldsymbol{x}), \tag{3.19}$$

where $\lambda_i$ are the Lagrange multipliers and $\mu > 0$ is the *penalty parameter*.

The following result shows that when we have knowledge of the exact Lagrange multiplier vector $\boldsymbol{\lambda}^*$, the solution $\boldsymbol{x}^*$ of (3.15) is a strict minimizer of $L_A(\boldsymbol{x}, \boldsymbol{\lambda}; \mu)$ for all $\mu$ sufficiently large.

**Theorem 3.6.** *Let* $\boldsymbol{x}^*$ *be a local solution of* (3.15) *and also a regular point for the constraints. If the second-order sufficient conditions specified in Theorem* 3.3 *are satisfied for* $\boldsymbol{\lambda} = \boldsymbol{\lambda}^*$, *then there is a threshold value* $\bar{\mu}$ *such that for all* $\mu \geqslant \bar{\mu}$, $\boldsymbol{x}^*$ *is a strict local minimizer of* $L_A(\boldsymbol{x}, \boldsymbol{\lambda}; \mu)$.

Now, consider the inequality-constrained problem

$$\min_{\boldsymbol{x}\in\mathbb{R}^n} \; f(\boldsymbol{x}) \tag{3.20a}$$

$$\text{s.t.} \quad c_i(\boldsymbol{x}) \geqslant 0, \quad i \in \mathcal{I}. \tag{3.20b}$$

It is possible to convert (3.21) into an equality-constrained problem by introducing a vector of additional variables $\boldsymbol{z} := \left(z_1, z_2, \ldots, z_{|\mathcal{I}|}\right)^{\top}$. This problem is given by

$$\min_{\boldsymbol{x}\in\mathbb{R}^n} \; f(\boldsymbol{x}) \tag{3.21a}$$

$$\text{s.t.} \quad c_i(\boldsymbol{x}) - z_i^2 = 0, \quad i \in \mathcal{I}. \tag{3.21b}$$

Note that $\boldsymbol{x}^*$ is a solution of (3.20) if and only if $(\boldsymbol{x}^*, \boldsymbol{z}^*)$, where $z_i^* := \sqrt{c_i(\boldsymbol{x}^*)}$, $i \in \mathcal{I}$, is a solution of problem (3.21).

Consider first the augmented Lagrangian for problem (3.21) defined for

$$\tilde{L}_A(\boldsymbol{x}, \boldsymbol{z}, \boldsymbol{\lambda}; \mu) = f(\boldsymbol{x}) - \sum_{i \in \mathcal{I}} \lambda_i \left(c_i(\boldsymbol{x}) - z_i^2\right) + \frac{\mu}{2} \sum_{i \in \mathcal{I}} \left(c_i(\boldsymbol{x}) - z_i^2\right)^2. \tag{3.22}$$

In applying the augmented Lagrangian method only involving inequality constraints, we must minimize the augmented Lagrangian (3.22) with respect to $(\boldsymbol{x}, \boldsymbol{z})$ for various values of $\boldsymbol{\lambda}$ and $\mu$. An important point here is that minimization of $\tilde{L}_A(\boldsymbol{x}, \boldsymbol{z}, \boldsymbol{\lambda}; \mu)$ with respect to $\boldsymbol{z}$ can be carried out explicitly for each fixed $\boldsymbol{x}$. To see this, note that

$$\min_{\boldsymbol{z}} \tilde{L}_A(\boldsymbol{x}, \boldsymbol{z}, \boldsymbol{\lambda}; \mu) = \min_{\boldsymbol{z}} \left\{ f(\boldsymbol{x}) - \sum_{i \in \mathcal{I}} \lambda_i (c_i(\boldsymbol{x}) - s_i) + \frac{\mu}{2} \sum_{i \in \mathcal{I}} (c_i(\boldsymbol{x}) - s_i)^2 \right\}. \tag{3.23}$$

The minimization with respect to $\boldsymbol{z}$ is equivalent to

$$\min_{\boldsymbol{s} \geq 0} \left\{ - \sum_{i \in \mathcal{I}} \lambda_i (c_i(\boldsymbol{x}) - s_i) + \frac{\mu}{2} \sum_{i \in \mathcal{I}} (c_i(\boldsymbol{x}) - s_i)^2 \right\}. \tag{3.24}$$

where $\boldsymbol{s} := \left(z_1^2, z_2^2, \ldots, z_{|\mathcal{I}|}^2\right)^\top$. The function in braces of (3.24) is quadratic in $s_i$. Its unconstrained solution is the $\boldsymbol{s}$ at which the derivative is zero. We have

$$s_i = \max \left\{ 0, c_i(\boldsymbol{x}) - \frac{1}{\mu} \lambda_i \right\}, \tag{3.25}$$

and thus

$$c_i(\boldsymbol{x}) - s_i = \min \left\{ c_i(\boldsymbol{x}), \frac{1}{\mu} \lambda_i \right\}. \tag{3.26}$$

As a result,

$$\begin{aligned} &\min_{\boldsymbol{s} \geq 0} \left\{ -\lambda_i (c_i(\boldsymbol{x}) - s_i) + \frac{\mu}{2} (c_i(\boldsymbol{x}) - s_i)^2 \right\} \\ &= \begin{cases} -\lambda_i c_i(\boldsymbol{x}) + \dfrac{\mu}{2} c_i^2(\boldsymbol{x}), & \text{if } c_i(\boldsymbol{x}) \leqslant \dfrac{1}{\mu} \lambda_i, \\ -\dfrac{\lambda_i^2}{2\mu}, & \text{otherwise.} \end{cases} \end{aligned} \tag{3.27}$$

Let us use the notation

$$\psi(c_i(\boldsymbol{x}), \lambda_i, \mu) := \begin{cases} -\lambda_i c_i(\boldsymbol{x}) + \dfrac{\mu}{2} c_i^2(\boldsymbol{x}), & \text{if } c_i(\boldsymbol{x}) \leqslant \dfrac{1}{\mu}\lambda_i, \\ -\dfrac{\lambda_i^2}{2\mu}, & \text{otherwise.} \end{cases} \tag{3.28}$$

We are thus led to the following definition of the augmented Lagrangian for (3.20)

$$\tilde{L}_A(\boldsymbol{x}, \boldsymbol{\lambda}; \mu) = f(\boldsymbol{x}) + \sum_{i \in \mathcal{I}} \psi(c_i(\boldsymbol{x}), \lambda_i; \mu). \tag{3.29}$$

An alternative expression for $\tilde{L}_A(\boldsymbol{x}, \boldsymbol{\lambda}; \mu)$ is given by

$$\tilde{L}_A(\boldsymbol{x}, \boldsymbol{\lambda}; \mu) = f(\boldsymbol{x}) + \frac{\mu}{2} \sum_{i \in \mathcal{I}} \left( \min^2 \left\{ 0, c_i(\boldsymbol{x}) - \frac{\lambda_i}{\mu} \right\} - \frac{\lambda_i^2}{\mu^2} \right). \tag{3.30}$$

The equality of the expressions (3.29) and (3.30) can be verified by a straightforward calculation.

The conclusion from the preceding discussion is that the problem

$$\min_{(\boldsymbol{x}, z) \in \mathbb{R}^{n+|\mathcal{I}|}} \tilde{L}_A(\boldsymbol{x}, z, \boldsymbol{\lambda}; \mu) \tag{3.31}$$

is equivalent to the problem

$$\min_{\boldsymbol{x} \in \mathbb{R}^n} \tilde{L}_A(\boldsymbol{x}, \boldsymbol{\lambda}; \mu) \tag{3.32}$$

and $(\boldsymbol{x}^*, z^*)$ is a solution of problem (3.31) if only if $\boldsymbol{x}^*$ is a solution of problem (3.32) and

$$z_i^{*2} = \max \left\{ 0, c_i(\boldsymbol{x}^*) - \frac{1}{\mu}\lambda_i \right\}. \tag{3.33}$$

As a result, the augmented Lagrangian methods of equality-constrained problem can be applied to inequality-constrained problem after it has been converted to the equality-constrained problem (3.21), but the computation itself need not involve the additional variables $z_1, z_2, \ldots, z_{|\mathcal{I}|}$ since we can solve the equivalent problem (3.32) in place of problem (3.31).

### *3.2.4 Sequential Quadratic Programming*

SQP is one of the most successful methods for the numerical solution of constrained nonlinear optimization problems. It relies on a profound theoretical foundation and provides powerful algorithmic tools for the solution of small and medium-size constrained optimization problems. The theory was initiated by Wilson in [263] and was further developed by Han in [96, 97] and Powell in [203].

We begin by considering the equality-constrained problem (3.15). The idea behind the SQP approach is to model (3.15) at the current iterate $\boldsymbol{x}_k$ by a quadratic programming subproblem and then use the minimizer of this subproblem to define a new iterate $\boldsymbol{x}_{k+1}$. The challenge is to design the quadratic subproblem so that it yields a good step for the nonlinear optimization problem.

From (3.9), we know that the Lagrangian function for this problem is $L(\boldsymbol{x}, \boldsymbol{\lambda}) = f(\boldsymbol{x}) - \sum_{i \in \mathcal{E}} \lambda_i c_i(\boldsymbol{x})$. We use $\boldsymbol{A}(\boldsymbol{x})$ to denote the Jacobian matrix of the constraints, that is,

$$\boldsymbol{A}(\boldsymbol{x}) = (\nabla c_1(\boldsymbol{x}), \nabla c_2(\boldsymbol{x}), \ldots, \nabla c_{|\mathcal{E}|}(\boldsymbol{x}))^\top. \tag{3.34}$$

The first-order KKT conditions (3.10) of the equality-constrained problem (3.15) can be written as a system of $n + |\mathcal{E}|$ equations in the $n + |\mathcal{E}|$ unknowns $\boldsymbol{x}$ and $\boldsymbol{\lambda}$

$$\boldsymbol{F}(\boldsymbol{x}, \boldsymbol{\lambda}) = \begin{pmatrix} \nabla f(\boldsymbol{x}) - \boldsymbol{A}(\boldsymbol{x})^\top \boldsymbol{\lambda} \\ \boldsymbol{c}(\boldsymbol{x}) \end{pmatrix} = \boldsymbol{0}. \tag{3.35}$$

Any solution $(\boldsymbol{x}^*, \boldsymbol{\lambda}^*)$ of the equality-constrained problem (3.15) for which $\boldsymbol{A}(\boldsymbol{x}^*)$ has full rank satisfies (3.35). One approach that suggests itself is to solve the nonlinear equations (3.35) by using Newton's method.

The Jacobian of (3.35) with respect to $\boldsymbol{x}$ and $\boldsymbol{\lambda}$ is given by

$$\boldsymbol{F}'(\boldsymbol{x}, \boldsymbol{\lambda}) = \begin{pmatrix} \nabla^2_{xx} L(\boldsymbol{x}, \boldsymbol{\lambda}) & -\boldsymbol{A}(\boldsymbol{x})^\top \\ \boldsymbol{A}(\boldsymbol{x}) & \boldsymbol{0} \end{pmatrix}. \tag{3.36}$$

The Newton step from the iterate $(\boldsymbol{x}_k, \boldsymbol{\lambda}_k)$ is thus given by

$$\begin{pmatrix} \boldsymbol{x}_{k+1} \\ \boldsymbol{\lambda}_{k+1} \end{pmatrix} = \begin{pmatrix} \boldsymbol{x}_k \\ \boldsymbol{\lambda}_k \end{pmatrix} + \begin{pmatrix} \boldsymbol{p}_k \\ \boldsymbol{p}_\lambda \end{pmatrix}, \tag{3.37}$$

where $\boldsymbol{p}_k$ and $\boldsymbol{p}_\lambda$ solve the Newton–KKT system

$$\begin{pmatrix} \nabla^2_{xx} L_k & -\boldsymbol{A}_k^\top \\ \boldsymbol{A}_k & \boldsymbol{0} \end{pmatrix} \begin{pmatrix} \boldsymbol{p}_k \\ \boldsymbol{p}_\lambda \end{pmatrix} = \begin{pmatrix} -\nabla f_k + \boldsymbol{A}_k^\top \boldsymbol{\lambda}_k \\ -\boldsymbol{c}_k \end{pmatrix}. \tag{3.38}$$

This Newton iteration is well defined when the KKT matrix in (3.38) is nonsingular. In fact, this matrix is nonsingular if the following assumption holds at $(\boldsymbol{x}, \boldsymbol{\lambda}) = (\boldsymbol{x}_k, \boldsymbol{\lambda}_k)$.

**Assumption 3.1.** (a) *The constraint Jacobian* $\boldsymbol{A}(\boldsymbol{x})$ *has full row rank;*
(b) *The matrix* $\nabla^2_{xx}L(\boldsymbol{x}, \boldsymbol{\lambda})$ *satisfies that* $\boldsymbol{d}^\top \nabla^2_{xx}L(\boldsymbol{x}, \boldsymbol{\lambda})\boldsymbol{d} > 0$ *for all* $\boldsymbol{d} \neq \boldsymbol{0}$ *such that* $\boldsymbol{A}^\top(\boldsymbol{x})\boldsymbol{d} = \boldsymbol{0}$.

There is an alternative way to view the iterations (3.37) and (3.38). Suppose that at the iterate $(\boldsymbol{x}_k, \boldsymbol{\lambda}_k)$ we model problem (3.15) using the quadratic program

$$\min_{p} \; f_k + \nabla f_k^\top \boldsymbol{p} + \frac{1}{2}\boldsymbol{p}^\top \nabla^2_{xx}L_k \boldsymbol{p} \tag{3.39a}$$

$$\text{s.t.} \quad \boldsymbol{A}_k \boldsymbol{p} + \boldsymbol{c}_k = \boldsymbol{0}. \tag{3.39b}$$

If Assumption 3.1 holds, this problem has a unique solution $(\boldsymbol{p}_k, \boldsymbol{l}_k)$ that satisfies

$$\nabla^2_{xx}L_k \boldsymbol{p}_k + \nabla f_k - \boldsymbol{A}_k^\top \boldsymbol{l}_k = \boldsymbol{0}, \tag{3.40a}$$

$$\boldsymbol{A}_k \boldsymbol{p}_k + \boldsymbol{c}_k = \boldsymbol{0}. \tag{3.40b}$$

The vectors $\boldsymbol{p}_k$ and $\boldsymbol{l}_k$ can be identified with the solution of the Newton equations (3.38). If we subtract $\boldsymbol{A}_k^\top \boldsymbol{\lambda}_k$ from both sides of the first equation in (3.38), we obtain

$$\begin{pmatrix} \nabla^2_{xx}L_k & -\boldsymbol{A}_k^\top \\ \boldsymbol{A}_k & \boldsymbol{0} \end{pmatrix} \begin{pmatrix} \boldsymbol{p}_k \\ \boldsymbol{\lambda}_{k+1} \end{pmatrix} = \begin{pmatrix} -\nabla f_k \\ -\boldsymbol{c}_k \end{pmatrix}. \tag{3.41}$$

Hence, by nonsingularity of the coefficient matrix, we have that $\boldsymbol{\lambda}_{k+1} = \boldsymbol{l}_k$ and that $\boldsymbol{p}_k$ solves (3.39) and (3.38).

The new iterate $(\boldsymbol{x}_{k+1}, \boldsymbol{\lambda}_{k+1})$ can therefore be defined either as the solution of the quadratic program (3.39) or as the iterate generated by Newton's method (3.37), (3.38) applied to the optimality conditions of the problem.

The above method can be extended to the general constrained mathematical problem (3.1)–(3.3). To model this problem we now linearize both the inequality and equality constraints to obtain

$$\min_{p} \; f_k + \nabla f_k^\top \boldsymbol{p} + \frac{1}{2}\boldsymbol{p}^\top \nabla^2_{xx}L_k \boldsymbol{p} \tag{3.42a}$$

$$\text{s.t.} \quad \nabla c_i(\boldsymbol{x}_k)^\top \boldsymbol{p} + c_i(\boldsymbol{x}_k) = \boldsymbol{0}, \quad i \in \mathcal{E}, \tag{3.42b}$$

$$\nabla c_i(\boldsymbol{x}_k)^\top \boldsymbol{p} + c_i(\boldsymbol{x}_k) \geq \boldsymbol{0}, \quad i \in \mathcal{I}. \tag{3.42c}$$

We can use one of the algorithms for quadratic programming to solve this problem. The new iterate is given by $(\boldsymbol{x}_k + \boldsymbol{p}_k, \boldsymbol{\lambda}_{k+1})$ where $\boldsymbol{p}_k$ and $\boldsymbol{\lambda}_{k+1}$ are the solution and the corresponding Lagrange multiplier of (3.42). A SQP method for (3.1)–(3.3) is thus given with the modification that the step is computed from (3.42).

In this inequality-constrained quadratic program approach, the set of active constraints $\mathcal{A}_k$ at the solution of (3.42) constitutes our guess of the active set at the solution of the nonlinear program. If the SQP method is able to correctly identify this optimal active set (and not change its guess at a subsequent iteration), then it will act like a Newton method for equality-constrained optimization and will converge rapidly.

## 3.3 Evolutionary Algorithms

Evolutionary optimization algorithms have become very popular in the last decade or two. Unlike the gradient-based techniques, where a single point is updated (typically using gradient information) from one iteration to the next, these algorithms do not require any gradient information and typically make use of a set of points to find the optima. These methods are typically inspired by some phenomena from nature and have the advantage of being extremely robust, having an increased chance of finding a global or near global optimum and being easy to implement. The big drawbacks associated with these algorithms are high computational cost, poor constraint-handling abilities, problem-specific parameter tuning, and limited problem size. In this section, two of the popular evolutionary algorithms, i.e., particle swarm optimization (PSO) and differential evolution (DE), and some constraint-handling techniques will be briefly introduced.

### *3.3.1 Particle Swarm Optimization*

PSO is based on a simplified social model [117,118,167,198,228]. PSO method was proposed by Kennedy [117] and has attracted considerable attention as one of the promising optimization methods with higher speed and higher accuracy than those of existing solution methods.

PSO is based on the social behavior that a population of individuals adapts to its environment by returning to promising regions that were previously discovered. This adaptation to the environment is a stochastic process that depends on both the memory of each individual, called *particle*, and the knowledge gained by the population, called *swarm*.

In the numerical implementation of this simplified social model, each particle has four attributes: the position vector in the search space, the velocity vector, the best position in its track, and the best position of the swarm. The process can be outlined as follows:

**Step 1.** Generate the initial swarm involving $N$ particles at random.
**Step 2.** Calculate the new velocity vector of each particle, based on its attributes.
**Step 3.** Calculate the new position of each particle from the current position and its new velocity vector.
**Step 4.** If the termination condition is satisfied, then stop. Otherwise, go to Step 2.

To be more specific, the new velocity vector of the $j$th particle at step $k+1$, $\boldsymbol{v}_{k+1}^j$, is calculated by the following scheme

$$\boldsymbol{v}_{k+1}^j := \omega_k \boldsymbol{v}_k^j + c_1 r_k^1 \cdot \left( \boldsymbol{pb}_k^j - \boldsymbol{x}_k^j \right) + c_2 r_k^2 \cdot \left( \boldsymbol{pg}_k - \boldsymbol{x}_k^i \right). \tag{3.43}$$

In (3.43), $r_k^1$ and $r_k^2$ are random numbers between 0 and 1, $\boldsymbol{pb}_k^j$ is the best position of the $j$th particle in its track, and $\boldsymbol{pg}_k$ is the best position of the swarm. There are three problem-dependent parameters: the inertia of the particle $\omega_k$ and two trust parameters $c_1$, $c_2$.

Then, the new position of the $j$th particle at step $k+1$, $\boldsymbol{x}_{k+1}^j$, is calculated by

$$\boldsymbol{x}_{k+1}^j := \boldsymbol{x}_k^j + \boldsymbol{v}_{k+1}^j, \tag{3.44}$$

where $\boldsymbol{x}_k^j$ is the current position of the $j$th particle at step $k$. The $j$th particle calculates the next search direction vector $\boldsymbol{v}_{k+1}^j$ by (3.43) in consideration of the current search direction vector $\boldsymbol{v}_k^j$, the direction vector going from the current search position $\boldsymbol{x}_k^j$ to the best position in its track $\boldsymbol{pb}_k^j$, and the direction vector going from the current search position $\boldsymbol{x}_k^j$ to the best position of the swarm $\boldsymbol{pg}_k$, and it moves from the current position $\boldsymbol{x}_k^j$ to the next search position $\boldsymbol{x}_{k+1}^j$ calculated by (3.44). The parameter $\omega_k$ controls the amount of the move by searching globally in the early stage and searching locally by decreasing $\omega_k$ gradually. It is defined by

$$\omega_k := \omega_0 - \frac{(\omega_0 - \omega_{T_{\max}})k}{T_{\max}}, \tag{3.45}$$

where $T_{\max}$ is the number of maximum iteration times, $\omega_0$ is an initial value at the time iteration, and $\omega_{T_{\max}}$ is the last value.

The values of $\boldsymbol{pb}_k^j$ and $\boldsymbol{gb}_k$ are updated as follows. Comparing the evaluation value of a particle after move, $f\left(\boldsymbol{x}_{k+1}^j\right)$, with that of the best position in its track, $f\left(\boldsymbol{pb}_k^j\right)$, if $f\left(\boldsymbol{x}_{k+1}^j\right)$ is better than $f\left(\boldsymbol{pb}_k^j\right)$, then the best position in its track is updated as $\boldsymbol{pb}_{k+1}^j := \boldsymbol{x}_{k+1}^j$. Furthermore, if $f\left(\boldsymbol{pb}_{k+1}^j\right)$ is better than $f(\boldsymbol{pg}_k)$, then the best position in the swarm is updated as $\boldsymbol{pg}_{k+1} := \boldsymbol{pb}_{k+1}^j$.

### 3.3.2 *Differential Evolution*

Similar to other evolutionary algorithms, DE is a population-based, derivative-free function optimizer. It usually encodes decision variables as floating-point numbers and manipulates them with simple arithmetic operations such as addition, subtraction, and multiplication. DE algorithm, proposed by Storn and Price [235], is a simple yet powerful population-based stochastic search technique, which is an efficient and effective global optimizer in the continuous search domain [204].

The initial population

$$\boldsymbol{x}_0^j := \left(x_{1,0}^j, x_{2,0}^j, \ldots, x_{n,0}^j\right)^{\top}, \quad j = 1, 2, \ldots, N \tag{3.46}$$

is randomly generated according to a normal or uniform distribution, where $N$ is the population size. After initialization, DE enters a loop of evolutionary operations: mutation, crossover, and selection.

*Mutation*: At step $k$, this operation creates mutation vectors $\boldsymbol{v}_k^j$ based on the current population

$$\boldsymbol{x}_k^j := \left(x_{1,k}^j, x_{2,k}^j, \ldots, x_{n,k}^j\right)^{\top}, \quad j = 1, 2, \ldots, N. \tag{3.47}$$

The followings are different mutation strategies frequently used in the literature:

- DE1

$$\boldsymbol{v}_k^j = \boldsymbol{x}_k^{r_0} + F \cdot \left(\boldsymbol{x}_k^{r_1} - \boldsymbol{x}_k^{r_2}\right), \tag{3.48}$$

- DE2

$$\boldsymbol{v}_k^j = \boldsymbol{x}_k^j + F \cdot \left(\boldsymbol{x}_k^b - \boldsymbol{x}_k^j\right) + F \cdot \left(\boldsymbol{x}_k^{r_1} - \boldsymbol{x}_k^{r_2}\right), \tag{3.49}$$

- DE3

$$\boldsymbol{v}_k^j = \boldsymbol{x}_k^b + F \cdot \left(\boldsymbol{x}_k^{r_1} - \boldsymbol{x}_k^{r_2}\right), \tag{3.50}$$

where the indices $r_0, r_1$ and $r_2$ are distinct integers uniformly chosen from the set $\{1, 2, \ldots, N\} \setminus \{j\}$; $\boldsymbol{x}_k^{r_1} - \boldsymbol{x}_k^{r_2}$ is a difference vector to mutate the population; $\boldsymbol{x}_k^b$ is the best vector at the current step $k$; and $F$ is the mutation factor which usually ranges on the interval $(0, 1]$.

The above mutation strategies can be generalized by implementing multiple difference vectors other than $\boldsymbol{x}_k^{r_1} - \boldsymbol{x}_k^{r_2}$.

*Crossover*: After mutation operation, a crossover operation forms the final trial vector $\boldsymbol{u}_k^j := \left(u_{1,k}^j, u_{2,k}^j, \ldots, u_{n,k}^j\right)^{\top}$:

$$\boldsymbol{u}_{l,k}^j = \begin{cases} \boldsymbol{v}_{l,k}^j, & \text{if } r^l \leqslant CR \text{ or } l = r_l^j, \\ \boldsymbol{x}_{l,k}^j, & \text{otherwise}, \end{cases} \tag{3.51}$$

where $r^l$ is a uniform random number on the interval $(0, 1]$ for each $l$; $r_l^j$ is an integer randomly chosen from 1 to $n$ for each $j$; and the crossover probability, $CR \in [0, 1]$, roughly corresponds to the average fraction of vector components that are inherited from the mutation vector.

*Selection*: The selection operation selects the better one from the vector $\boldsymbol{x}_k^j$ and the trial vector $\boldsymbol{u}_k^j$ according to their fitness values $f(\cdot)$. The selected vector is given by

$$\boldsymbol{x}_{k+1}^j := \begin{cases} \boldsymbol{u}_k^j, \text{ if } f\left(\boldsymbol{u}_k^j\right) < f\left(\boldsymbol{x}_k^j\right), \\ \boldsymbol{x}_k^j, \text{ otherwise,} \end{cases} \tag{3.52}$$

and used as vector in the next step.

The above one-to-one selection procedure is generally kept fixed in different DE algorithms, while the crossover may have variants other than the operation in (3.51).

### *3.3.3 Constraint-Handling Techniques*

A common characteristic of most of the evolutionary algorithms is inherently an unconstrained optimization algorithm. This is a major drawback associated with these methods. Many different constraint-handling techniques have been investigated to deal with this problem. Generally, constraint-handling techniques for evolutionary algorithms are classified as: techniques based on penalty functions, techniques that preserve feasibility, techniques making a clear distinction between feasible and infeasible solutions; and other hybrid techniques [126]. Among these, the most popular is to make use of a penalty function approach [20, 105, 112], and we will briefly introduce this approach in the sequel. For other constraint-handling techniques, the reader may refer to [218, 229].

In the sense of the penalty function method, the constrained problem is transformed into unconstrained by using the modified evaluation function

$$\tilde{Q}(\boldsymbol{x}) = \begin{cases} f(\boldsymbol{x}), & \text{if } \boldsymbol{x} \in \Omega, \\ f(\boldsymbol{x}) + p(\boldsymbol{x}), & \text{otherwise,} \end{cases} \tag{3.53}$$

where $p(\boldsymbol{x})$ is zero if no violation occurs and is positive, otherwise. The penalty function is usually based on some form of distance that measures how far the infeasible solution is from the feasible region $\Omega$ or on the effort to "repair" the solution, i.e., to transform it into $\Omega$. Many methods rely on a set of function $f_i(\boldsymbol{x})$, $i \in \mathcal{E} \cup \mathcal{I}$, to construct the penalty, where the function $f_i$ measures the violation of the $i$th constraint

$$f_i(\boldsymbol{x}) = \begin{cases} |c_i(\boldsymbol{x})|, & i \in \mathcal{E}, \\ \min\{0, c_i(\boldsymbol{x})\}, & i \in \mathcal{I}. \end{cases} \tag{3.54}$$

But these methods differ in many important details with respect to how the penalty function is designed and applied to infeasible solutions. Some more details on the specifics is provided below.

A *static penalty function* method uses the following equation to evaluate each particle

$$\tilde{Q}(x) = f(\boldsymbol{x}) + \sum_{i \in \mathcal{E} \cup \mathcal{I}} r_i f_i^2(\boldsymbol{x}), \tag{3.55}$$

where $r_i$ are the penalty parameters. The weakness of the method arises in the number of the penalty parameters. In contrast to the static penalty function method, a *dynamic penalty* method arises and individuals are evaluated at step $k$, by the formula

$$\tilde{Q}(\boldsymbol{x}) = f(\boldsymbol{x}) + (c \times k)^{\alpha} \sum_{i \in \mathcal{E} \cup \mathcal{I}} f_i^{\beta}(\boldsymbol{x}), \tag{3.56}$$

where $c > 0$, $\alpha > 0$, and $\beta > 0$ are constants. This method doesn't require as many parameters as the static method, and instead of defining several violation levels, the selection pressure on infeasible solutions increases due to the $(c \times k)^{\alpha}$ component of the penalty term: as $k$ grows larger, this component also grows larger. An *adaptive penalty* method where the penalty function takes a feedback from the search process is developed and each particle is evaluated by the formula

$$\tilde{Q}(\boldsymbol{x}) = f(\boldsymbol{x}) + r_k \cdot \sum_{i \in \mathcal{E} \cup \mathcal{I}} f_i^2(\boldsymbol{x}), \tag{3.57}$$

where $r_k$ is updated at step $k$ using

$$r_{k+1} := \begin{cases} \dfrac{r_k}{\beta_1}, & \text{if } \boldsymbol{pg}_{\ell} \in \Omega \text{ for all } k - l + 1 \leqslant \ell \leqslant k, \\ \beta_2 r_k & \text{if } \boldsymbol{pg}_{\ell} \in \mathbb{R}^n \setminus \Omega \text{ for all } k - l + 1 \leqslant \ell \leqslant k, \\ r_k, & \text{otherwise,} \end{cases} \tag{3.58}$$

where $\boldsymbol{pg}_{\ell}$ denotes the best individual in terms of function evaluation $\tilde{Q}$ in (3.57) at step $\ell$, $\beta_1, \beta_2 > 1$, and $\beta_1 \neq \beta_2$ to avoid cycling. In other words, (i) if all of the best individuals in the last $l$ steps were feasible, then the method decreases the penalty component $r_{k+1}$ for step $k + 1$; (ii) if all of the best individuals in the last $l$ steps were infeasible, then the method increases the penalties; and (iii) if there are some feasible and infeasible individuals as best individuals in the last $k$ steps, then $r_{k+1}$ remains without change.

# Chapter 4
# Elements of Optimal Control Theory

## 4.1 Introduction

Optimal control deals with the problem of finding a control law for a given system such that a certain optimality criterion is achieved. There are already many excellent books devoted solely to the detailed exposition of the theory of optimal control. We refer the interested reader to [1,7,29,59,165,192] for systems described by ordinary differential equations, delayed differential equations, switched systems, and partial differential equations, respectively.

In the next section, we present three classes of dynamical systems, i.e., ordinary differential system, delay-differential system, and switched system, for which the optimal control problems are concerned in this book. Some basic concepts and theorems for these dynamical systems are briefly introduced. In the following sections, we focus on briefly introducing the optimal control theory for systems described by ordinary differential equations. In Sect. 4.3, we briefly formulate the standard optimal control problem and the optimal multiprocess control problem. In Sect. 4.4, the necessary conditions for the standard optimal control problem and optimal multiprocesses are briefly discussed. For results on the existence of optimal controls and sufficient conditions, we refer the interested reader to [7,8,42,166,169].

## 4.2 Dynamical Systems

### *4.2.1 Ordinary Differential System*

Any equation containing differential coefficients is called a *differential equation*. *Ordinary differential equations* are those that involve only one independent variable and therefore only ordinary differential coefficients. Our main task in this section will be to introduce some basic results for ordinary differential equations.



C. Liu, Z. Gong, *Optimal Control of Switched Systems Arising in Fermentation Processes*, Springer Optimization and Its Applications 97,
DOI 10.1007/978-3-662-43793-3_4

Let $G \subseteq \mathbb{R}^m$, $H \subseteq \mathbb{R}^n$, and $k \in \mathbb{N}_0 = \mathbb{N} \cup \{0\}$. Then $C^k(G, H)$ denotes the set of functions $G \to H$ having continuous derivatives up to order $k$. In addition, we will abbreviate $C^k(G) = C^k(G, \mathbb{R})$ and $C(G, H) = C^0(G, H)$ to denote the set of continuous functions from $G$ to $H$.

A classical *k-order ordinary differential equation* (ODE) is a relation of the form

$$F\left(t, x, x^{(1)}, \cdots, x^{(k)}\right) = 0 \tag{4.1}$$

for the unknown function $x \in C^k(I), I \subseteq \mathbb{R}$. Here $F \in C(G)$ with $G$ as an open subset of $\mathbb{R}^{k+2}$ and

$$x^{(k)}(t) := \frac{\mathrm{d}^k x(t)}{\mathrm{d}t^k}, \quad k \in \mathbb{N}_0, \tag{4.2}$$

are the *k-order ordinary* derivatives of $x$. One frequently calls $t$ the *independent* and $x$ the *dependent variable*. A *solution* of the ODE (4.1) is a function $\phi \in C^k(\bar{I})$, where $\bar{I} \subseteq I$ is an interval, such that

$$F\left(t, \phi(t), \phi^{(1)}(t), \ldots, \phi^{(k)}(t)\right) = 0, \quad \text{for all } t \in \bar{I}. \tag{4.3}$$

This implicitly implies $\left(t, \phi(t), \phi^{(1)}(t), \ldots, \phi^{(k)}(t)\right) \in G$ for all $t \in \bar{I}$.

Unfortunately there is not too much one can say about differential equations in the above form (4.1). Hence we will assume that one can solve $F$ for the highest derivative resulting in a differential equation of the form

$$x^{(k)} = f\left(t, x, x^{(1)}, \ldots, x^{(k-1)}\right). \tag{4.4}$$

By the implicit function theorem, this can be done at least locally near some point $(t, y) \in G$ if the partial derivative with respect to the highest derivative does not vanish at that point, $\frac{\partial F(t, y)}{\partial y_k}(t, y) \neq 0$. This is the type of differential equations we will from now on look at.

It should be noted that the case of real-valued functions is not enough and we should admit the case $\boldsymbol{x} : \mathbb{R} \to \mathbb{R}^n$. This leads us to *systems of ordinary differential equations*

$$\begin{aligned} x_1^{(k)} &= f_1\left(t, x, x^{(1)}, \ldots, x^{(k-1)}\right), \\ &\vdots \\ x_n^{(k)} &= f_n\left(t, x, x^{(1)}, \ldots, x^{(k-1)}\right). \end{aligned} \tag{4.5}$$

Moreover, any system can always be reduced to a first-order system by changing to the new set of independent variables $\boldsymbol{y} := \left(x, x^{(1)}, \ldots, x^{(k-1)}\right)^\top$. This yields the following *first-order system*

$$
\begin{aligned}
\dot{y}_1 &= y_2, \\
&\vdots \\
\dot{y}_{k-1} &= y_k, \\
\dot{y}_k &= f(t, \mathbf{y}).
\end{aligned}
\tag{4.6}
$$

Thus, it suffices to consider the case of the first-order systems.

Consider the following *initial value problem* (IVP)

$$
\begin{cases}
\dot{\mathbf{x}}(t) = \mathbf{f}(t, \mathbf{x}(t)), \\
\mathbf{x}(t_0) = \mathbf{x}_0.
\end{cases}
\tag{4.7}
$$

To establish the existence and uniqueness of the solution for the IVP (4.7), we need the following condition. Suppose $\mathbf{f}$ is locally *Lipschitz continuous* in the second argument. That is, for every compact set $H \subseteq G$, the following number

$$
K := \sup_{(t,x)\neq(t,y)\in H} \frac{\|\mathbf{f}(t,\mathbf{x}) - \mathbf{f}(t,\mathbf{y})\|}{\|\mathbf{x} - \mathbf{y}\|} < +\infty \tag{4.8}
$$

(which depends on $H$) is finite, where $\|\cdot\|$ denotes the Euclidean norm.

**Theorem 4.1 (Picard–Lindelöf).** *Suppose $\mathbf{f} \in C(G, \mathbb{R}^n)$, where $G$ is an open subset of $\mathbb{R}^{n+1}$ and $(t_0, \mathbf{x}_0) \in G$. If $\mathbf{f}$ is locally Lipschitz continuous in the second argument, then there exists a unique local solution, denoted by $\mathbf{x}(\cdot|t_0, \mathbf{x}_0)$, of the* IVP (4.7).

Usually, in applications several data are only known approximately. If the problem is well posed, one expects that small changes in the data will result in small changes of the solution. This will be shown in the next theorem. As a preparation we need *Gronwall's inequality*.

**Lemma 4.1 (Gronwall's Inequality).** *Suppose $\vartheta(t), \varphi(t) \in C\left(\tilde{I}\right), \tilde{I} := [t_0, t_f]$ and satisfy*

$$
\varphi(t) \leqslant \alpha + \int_{t_0}^{t} (\vartheta(\tau)\varphi(\tau) + \beta)\mathrm{d}\tau \tag{4.9}
$$

*with $\alpha, \beta, \vartheta(\tau) \geqslant 0$. Then*

$$
\varphi(t) \leqslant (\alpha + (t_f - t_0)\beta) \exp\left(\int_{t_0}^{t} \vartheta(\tau)\mathrm{d}\tau\right), \quad \forall t \in \tilde{I}. \tag{4.10}
$$

Now, we can show that the IVP is well posed.

**Theorem 4.2.** *Suppose $\mathbf{f}, \mathbf{g} \in C(G, \mathbb{R}^n)$ and let $\mathbf{f}$ be Lipschitz continuous with constant $K$. If $\mathbf{x}(\cdot|t_0, \mathbf{x}_0)$ and $\mathbf{y}(\cdot|t_0, \mathbf{y}_0)$ are the respective solutions of the* IVPs

$$\begin{cases} \dot{\boldsymbol{x}}(t) = \boldsymbol{f}(t, \boldsymbol{x}(t)), \\ \boldsymbol{x}(t_0) = \boldsymbol{x}_0, \end{cases} \tag{4.11}$$

*and*

$$\begin{cases} \dot{\boldsymbol{y}}(t) = \boldsymbol{g}(t, \boldsymbol{y}(t)), \\ \boldsymbol{y}(t_0) = \boldsymbol{y}_0, \end{cases} \tag{4.12}$$

*then*

$$\|\boldsymbol{x}(t|t_0, \boldsymbol{x}_0) - \boldsymbol{y}(t|t_0, \boldsymbol{y}_0)\| \leqslant \|\boldsymbol{x}_0 - \boldsymbol{y}_0\| \exp(K|t - t_0|) + \frac{M}{K}(\exp(K|t - t_0|) - 1), \tag{4.13}$$

*where*

$$M := \sup_{(t,\boldsymbol{x}) \in G} \|\boldsymbol{f}(t, \boldsymbol{x}) - \boldsymbol{g}(t, \boldsymbol{x})\|. \tag{4.14}$$

The next theorem is to show that solutions exist for all $t \in \mathbb{R}$ if $\boldsymbol{f}(t, \boldsymbol{x})$ grows at most linearly with respect to $\boldsymbol{x}$.

**Theorem 4.3.** *Suppose* $G = \mathbb{R} \times \mathbb{R}^n$ *and there is a constant* $K > 0$ *such that*

$$\|\boldsymbol{f}(t, \boldsymbol{x})\| \leqslant K(1 + \|\boldsymbol{x}\|). \tag{4.15}$$

*Then all solutions of the* IVP (4.7) *are defined for all* $t \in \mathbb{R}$.

In fact, we can also handle the dependence on parameters. Suppose $\boldsymbol{f}$ depends on some parameters $\boldsymbol{\lambda} \in \Lambda \subseteq \mathbb{R}^p$ and consider the IVP

$$\begin{cases} \dot{\boldsymbol{x}}(t) = \boldsymbol{f}(t, \boldsymbol{x}(t), \boldsymbol{\lambda}), \\ \boldsymbol{x}(t_0) = \boldsymbol{x}_0, \end{cases} \tag{4.16}$$

with corresponding solution $\boldsymbol{x}(\cdot|t_0, \boldsymbol{x}_0, \boldsymbol{\lambda})$.

**Theorem 4.4.** *Suppose* $\boldsymbol{f} \in C(G \times \Lambda, \mathbb{R}^n)$ *and* $\boldsymbol{f}$ *is locally Lipschitz continuous in* $\boldsymbol{x}$. *Around each point* $(t_0, \boldsymbol{x}_0, \boldsymbol{\lambda}_0) \in G \times \Lambda$, *we can find an open set* $I_0 \times H_0 \times \Lambda_0 \subseteq G \times \Lambda$ *such that* $\boldsymbol{x}(\cdot|t_0, \boldsymbol{x}_0, \boldsymbol{\lambda})$ *is continuous in* $\boldsymbol{\lambda}$.

### 4.2.2 Delay-Differential System

Delay-differential equations differ from ordinary differential equations in that the derivative at any time depends on the solution at prior times. In this section, we shall introduce a mathematical framework and some basic properties for delay-differential equations without proofs. The main references are [70, 95].

For $r > 0$ and $H \subseteq \mathbb{R}^n$, let $C([-r,0],H)$ be the set of continuous functions mapping $[-r,0]$ into $H$.

**Definition 4.1.** If $\boldsymbol{x}$ is a function defined at least on $[t-r,t] \to \mathbb{R}^n$, then we define a new function $\boldsymbol{x}_t : [-r,0] \to \mathbb{R}^n$ by

$$\boldsymbol{x}_t(\theta) = \boldsymbol{x}(t+\theta), \quad -r \leqslant \theta \leqslant 0. \tag{4.17}$$

Clearly, if $\boldsymbol{x}$ is continuous on $[t-r,t] \to H \subseteq \mathbb{R}^n$, then $\boldsymbol{x}_t$ is continuous on $[-r,0]$. In the following, unless otherwise stated, we will take $I \subseteq \mathbb{R}$ and $H \subseteq \mathbb{R}^n$ to be open sets.

**Definition 4.2.** Let $\tilde{\boldsymbol{F}} : I \times C([-r,0],H) \to \mathbb{R}^n$ be a given functional. Then the relation

$$\dot{\boldsymbol{x}}(t) = \tilde{\boldsymbol{F}}(t,\boldsymbol{x}_t) \tag{4.18}$$

is called *delay-differential equations* (DDEs) on $I \times C([-r,0],H)$.

Note that (4.18) includes

(a) *Differential equations with discrete delays*:

$$\begin{aligned}\dot{\boldsymbol{x}}(t) &= \tilde{\boldsymbol{F}}(t,\boldsymbol{x}_t)\\ &= \boldsymbol{f}(t,\boldsymbol{x}_t(-\tau_1),\ldots,\boldsymbol{x}_t(-\tau_m))\\ &= \boldsymbol{f}(t,\boldsymbol{x}(t-\tau_1),\ldots,\boldsymbol{x}(t-\tau_m)).\end{aligned}$$

Here $\tau_j \geqslant 0$ is constant and $r := \max\limits_{1\leqslant j\leqslant m} \tau_j$.

(b) *Differential equations with bounded variable delays*:

$$\begin{aligned}\dot{\boldsymbol{x}}(t) &= \tilde{\boldsymbol{F}}(t,\boldsymbol{x}_t)\\ &= \boldsymbol{f}(t,\boldsymbol{x}_t(-\tau_1(t)),\ldots,\boldsymbol{x}_t(-\tau_m(t)))\\ &= \boldsymbol{f}(t,\boldsymbol{x}(t-\tau_1(t)),\ldots,\boldsymbol{x}(t-\tau_m(t))).\end{aligned}$$

Here $0 \leqslant \tau_j(t) \leqslant r$, $j = 1,\ldots,m$, $t \in I$.

(c) *Differential equations with a distribution of delays*:

$$\begin{aligned}\dot{\boldsymbol{x}}(t) &= \tilde{\boldsymbol{F}}(t,\boldsymbol{x}_t)\\ &= \int_{-r}^{0} \boldsymbol{f}(t,\theta,\boldsymbol{x}_t(\theta))\mathrm{d}\theta\\ &= \int_{-r}^{0} \boldsymbol{f}(t,\theta,\boldsymbol{x}(t+\theta))\mathrm{d}\theta.\end{aligned}$$

We may now give a precise definition of a solution of DDEs.

**Definition 4.3.** Let $\tilde{\boldsymbol{F}} : I \times C([-r,0],H) \to \mathbb{R}^n$. A function $\boldsymbol{x}(t)$ is said to be a *solution of* (4.18) *on* $[t_0 - r, \beta)$ if there are $t_0 \in \mathbb{R}$ and $\beta > t_0$ such that

(a) $\boldsymbol{x} \in C([t_0 - r, \beta), H)$;
(b) $[t_0, \beta) \subset I$; and
(c) $\boldsymbol{x}(t)$ satisfies (4.18) for $t \in [t_0, \beta)$.

For given $t_0 \in \mathbb{R}$ and $\boldsymbol{\phi}_0 \in C([-r,0],H)$, the initial value problem associated with the DDEs (4.18) is

$$\begin{cases} \dot{\boldsymbol{x}}(t) = \tilde{\boldsymbol{F}}(t, \boldsymbol{x}_t), & t \geqslant t_0, \\ \boldsymbol{x}(t) = \boldsymbol{\phi}_0(t - t_0), & t_0 - r \leqslant t \leqslant t_0. \end{cases} \tag{4.19}$$

**Definition 4.4.** The function $\boldsymbol{x}(t)$ is a *solution of the initial value problem* (4.19) *on* $[t_0 - r, \beta)$ if $\boldsymbol{x}(t)$ is a solution of (4.18) on $[t_0 - r, \beta)$ and $\boldsymbol{x}_{t_0} = \boldsymbol{\phi}_0$.

The following lemmas will be useful when discussing the properties of solutions.

**Lemma 4.2.** *If $\boldsymbol{x}$ is continuous on $[t_0 - r, t_0 + \gamma]$, then $\boldsymbol{x}_t$ is a continuous function of $t$ for $t \in [t_0, t_0 + \gamma]$.*

**Lemma 4.3.** *Let $\tilde{\boldsymbol{F}} : I \times C([-r,0],H) \to \mathbb{R}^n$ be continuous and let $t_0 \in I$ and $\boldsymbol{\phi}_0 \in C([-r,0],H)$ be given. Then $\boldsymbol{x}$ is a solution of the initial value problem* (4.19) *on $[t_0 - r, \beta)$ if and only if $[t_0, \beta) \subset I$, $\boldsymbol{x} \in C([t_0 - r, \beta), H)$ and $\boldsymbol{x}$ satisfies*

$$\begin{cases} \boldsymbol{x}(t) = \boldsymbol{\phi}_0(t - t_0), & t_0 - r \leqslant t \leqslant t_0, \\ \boldsymbol{x}(t) = \boldsymbol{\phi}_0(t_0) + \displaystyle\int_{t_0}^{t} \tilde{\boldsymbol{F}}(s, \boldsymbol{x}_s)\mathrm{d}s, & t_0 \leqslant t < \beta. \end{cases} \tag{4.20}$$

**Definition 4.5.** Let $\tilde{\boldsymbol{F}} : I \times C([-r,0],H) \to \mathbb{R}^n$ and let $\mathcal{E} \subset I \times C([-r,0],H)$. Then $F$ is *Lipschitz* on $\mathcal{E}$ if there exists a constant $K > 0$ such that

$$\left\| \tilde{\boldsymbol{F}}(t, \boldsymbol{\psi}) - \tilde{\boldsymbol{F}}\left(t, \bar{\boldsymbol{\psi}}\right) \right\| \leqslant K \left\| \boldsymbol{\psi} - \bar{\boldsymbol{\psi}} \right\|, \tag{4.21}$$

whenever $(t, \boldsymbol{\psi}), (t, \bar{\boldsymbol{\psi}}) \in \mathcal{E}$.

**Definition 4.6.** $\tilde{\boldsymbol{F}} : I \times C([-r,0],H) \to \mathbb{R}^n$ is *locally Lipschitz* if, for each given $(\bar{t}, \bar{\boldsymbol{\psi}}) \in I \times C([-r,0],H)$, there exist constants $a > 0$ and $b > 0$ such that

$$\mathcal{E} := ([\bar{t} - a, \bar{t} + a] \cap I) \times \{\boldsymbol{\psi} \in C([-r,0],\mathbb{R}^n) | \, \|\boldsymbol{\psi} - \bar{\boldsymbol{\psi}}\| \leqslant b\}$$

is a subset of $I \times C([-r,0],H)$ and $\tilde{\boldsymbol{F}}$ is Lipschitz on $\mathcal{E}$.

On the basis of the above definitions and lemmas, we now state the following important properties of solutions.

**Theorem 4.5 (Local Existence).** *Let $\tilde{\boldsymbol{F}} : [t_0, \alpha) \times C([-r, 0], H) \to \mathbb{R}^n$ be continuous and locally Lipschitz on its domain. Then, given any $\boldsymbol{\phi}_0 \in C([-r, 0], H)$ and $\beta \in (t_0, \alpha]$, there is at most one solution of the initial value problem* (4.19) *on* $[t_0 - r, \beta)$.

**Theorem 4.6 (Uniqueness).** *Let $\tilde{\boldsymbol{F}} : [t_0, \alpha) \times C([-r, 0], H) \to \mathbb{R}^n$ be continuous and locally Lipschitz. Then, for each $\boldsymbol{\phi}_0 \in C([-r, 0], H)$, the initial value problem* (4.19) *has a unique solution on $[t_0 - r, t_0 + \delta)$ for some $\delta > 0$.*

**Definition 4.7.** Let $\boldsymbol{x}$ on $[t_0 - r, \beta_1)$ and $\boldsymbol{y}$ on $[t_0 - r, \beta_2)$ be two solutions for the initial value problem (4.19). If $\beta_2 > \beta_1$ and $\boldsymbol{x}(t) = \boldsymbol{y}(t)$ for $t \in [t_0 - r, \beta_1)$, $\boldsymbol{y}$ is said to a *continuation* of $\boldsymbol{x}$ or $\boldsymbol{x}$ *can be continued* to $[t_0 - r, \beta_2)$. A solution $\boldsymbol{x}$ of (4.19) is *noncontinuable* if it has no continuation.

**Theorem 4.7 (Global Existence).** *Let $\tilde{\boldsymbol{F}} : [t_0, \alpha) \times C([-r, 0], \mathbb{R}^n) \to \mathbb{R}^n$ be continuous and locally Lipschitz. If*

$$\left\| \tilde{\boldsymbol{F}}(t, \boldsymbol{\psi}) \right\| \leqslant M(t) + N(t)\|\boldsymbol{\psi}\| \text{ on } [t_0, \alpha) \times C([-r, 0], \mathbb{R}^n), \tag{4.22}$$

*where $M$ and $N$ are continuous, positive functions on $[t_0, \alpha)$, then the unique noncontinuable solution of* (4.19) *exists on the entire interval $[t_0 - r, \alpha)$.*

**Theorem 4.8 (Continuous Dependence on Initial Conditions).** *Let $\tilde{\boldsymbol{F}} : [t_0, \alpha) \times C([-r, 0], H) \to \mathbb{R}^n$ be continuous and Lipschitz with Lipschitz constant $K$. Let $\boldsymbol{\phi}_0 \in C([-r, 0], H)$ and $\bar{\boldsymbol{\phi}}_0 \in C([-r, 0], H)$ be given, and let $\boldsymbol{x}$ and $\bar{\boldsymbol{x}}$ be unique solutions of* (4.19) *with $\boldsymbol{x}_{t_0} = \boldsymbol{\phi}_0$ and $\bar{\boldsymbol{x}}_{t_0} = \bar{\boldsymbol{\phi}}_0$, respectively. If $\boldsymbol{x}$ and $\bar{\boldsymbol{x}}$ are both valid on $[t_0 - r, \beta)$, then*

$$\|\boldsymbol{x}(t) - \bar{\boldsymbol{x}}(t)\| \leqslant \left\|\boldsymbol{\phi}_0 - \bar{\boldsymbol{\phi}}_0\right\| \exp\left(K(t - t_0)\right) \tag{4.23}$$

*for $t_0 \leqslant t < \beta$.*

**Theorem 4.9 (Continuous Dependence on $\tilde{\boldsymbol{F}}$).** *Let $\tilde{\boldsymbol{F}}, \bar{\boldsymbol{F}} : [t_0, \alpha) \times C([-r, 0], H) \to \mathbb{R}^n$ be continuous, and let $\bar{\boldsymbol{F}}$ be Lipschitz with Lipschitz constant $K$. Given $\boldsymbol{\phi}_0$, $\bar{\boldsymbol{\phi}}_0 \in C([-r, 0], H)$, let $\boldsymbol{x}(t)$ and $\bar{\boldsymbol{x}}(t)$ be the unique solutions of* (4.19) *and*

$$\begin{cases} \dot{\bar{\boldsymbol{x}}}(t) = \bar{\boldsymbol{F}}(t, \bar{\boldsymbol{x}}_t), & t \geqslant t_0, \\ \bar{\boldsymbol{x}}(t) = \bar{\boldsymbol{\phi}}_0(t - t_0), & t_0 - r \leqslant t \leqslant t_0, \end{cases} \tag{4.24}$$

*respectively. If $\boldsymbol{x}$ and $\bar{\boldsymbol{x}}$ are both valid on $[t_0 - r, \beta)$ and $\left\| \tilde{\boldsymbol{F}}(t, \boldsymbol{\psi}) - \bar{\boldsymbol{F}}(t, \boldsymbol{\psi}) \right\| \leqslant \mu$ for all $t \in [t_0, \alpha)$, $\boldsymbol{\psi} \in C([-r, 0], H)$, then*

$$\|\boldsymbol{x}(t) - \bar{\boldsymbol{x}}(t)\| \leqslant \left\|\boldsymbol{\phi}_0 - \bar{\boldsymbol{\phi}}_0\right\| \exp\left(K(t - t_0)\right) + \frac{\mu}{K}\left(\exp\left(K(t - t_0)\right) - 1\right) \tag{4.25}$$

*for $t_0 \leqslant t < \beta$.*

### 4.2.3 Switched System

Generally speaking, a switched system is a dynamical system in which switching plays a nontrivial role. The system admits continuous states that take values from a vector space and discrete states that take values from a discrete index set. The interaction between the continuous and discrete states makes switched systems widely representative and complicatedly behaved. In this section, we shall introduce some basic concepts for switched system. The main references are [59, 165, 274].

A *switched system* is mathematically described by

$$\dot{\boldsymbol{x}}(t) = \boldsymbol{f}^{i(t)}(t, \boldsymbol{x}(t), \boldsymbol{u}(t)), \tag{4.26a}$$

$$i(t) = \varphi(t, \boldsymbol{x}(t), i(t-)), \tag{4.26b}$$

where $\boldsymbol{x}(t) \in \mathbb{R}^n$ is the continuous state at time $t$; $i(t)$ is the discrete state at time $t$ taking value from an index set $P := \{1, 2, \ldots, \bar{P}\}$; $\boldsymbol{u}(t) \in \mathbb{R}^m$ is the control input; $\boldsymbol{f}^p : \mathbb{R} \times \mathbb{R}^n \times \mathbb{R}^m \to \mathbb{R}^n$, $p \in P$, is the vector field; and the function $\varphi : \mathbb{R} \times \mathbb{R}^n \times P \to P$ determines the active subsystem at time $t$. In general, $\varphi$ is not decided a priori but a part of the switching law.

In the system description, each individual mode

$$\dot{\boldsymbol{x}}(t) = \boldsymbol{f}^p(t, \boldsymbol{x}(t), \boldsymbol{u}(t)) \tag{4.27}$$

for $p \in P$ is said to be a *subsystem* of the switched system.

**Definition 4.8.** For a switched system, if each subsystem is autonomous (i.e., no continuous input), then it is called a *switched autonomous system*.

**Definition 4.9.** A switched system is called *time-dependent switched system* if the switching law depends only on time and is independent of the system's state.

**Definition 4.10.** A switched system is called *state-dependent switched system* if the mode switches occur when the system state intersects certain switching surfaces in the state space $\mathbb{R}^n$.

For a switched system, the presence of switching makes the behavior of the system more complicated than that of conventional systems. In particular, the evolution of the continuous and discrete states will leave us with a timed sequence of active subsystems that is defined as a switching sequence as follows.

**Definition 4.11.** A *switching sequence* $\sigma$ in $\tilde{I} = [t_0, t_f]$ is a timed sequence $\sigma = \left((t_0, i_0), (t_1, i_1), \ldots, (t_{\bar{K}}, i_{\bar{K}})\right)$, where $0 \leqslant \bar{K} \leqslant \infty$, $t_0 \leqslant t_1 \leqslant \cdots \leqslant t_{\bar{K}} \leqslant t_f$, and $i_k \in P$ for $0 \leqslant k \leqslant \bar{K}$.

A switching sequence $\sigma$ defined above indicates that the system starts from subsystem $i_0$ at $t_0$ and switches to subsystem $i_k$ from subsystem $i_{k-1}$ at $t_k$ for $1 \leqslant k \leqslant \bar{K}$. Subsystem $i_k$ will remain active in $[t_k, t_{k+1})$.

We give the solution concept of a switched system as follows.

**Definition 4.12.** For a given control input $\boldsymbol{u}$, a solution in $\tilde{I}$ to the switched system is any pair $(\boldsymbol{x}, i)$ with $i \in P$ and $\boldsymbol{x}$ a solution to

$$\dot{\boldsymbol{x}}(t) = \boldsymbol{f}^{i(t)}(t, \boldsymbol{x}(t), \boldsymbol{u}(t)), \quad t \in \tilde{I}.$$

**Definition 4.13.** A solution of switched system (4.26) is *regular* if $i(t)$ is piecewise constant with finitely many switching times.

Note that there is a peculiar type of behavior that can occur in switched systems, i.e., the discrete state takes infinitely many switchings in finite amount of time. This is so-called *Zeno behavior*. In this book, we either explicitly rule out Zeno behavior or show that it cannot occur.

Finally, we give the following concluding remarks for switched systems.

*Remark 4.1.* A switched system with time delays in the individual subsystem dynamics is called a *switched time-delay system*.

*Remark 4.2.* For a switched system, if the switching law $\sigma$ is decided a priori, then it degenerates into a multistage system.

## 4.3 Optimal Control Problems

As stated before, the optimal control problem is to determine the control policy that will cause a process to satisfy the physical constraints and at the same time minimize (or maximize) certain performance criterion.

### *4.3.1 Standard Optimal Control Problem*

Consider a single-stage process system described by the ordinary differential equations

$$\dot{\boldsymbol{x}}(t) = \boldsymbol{f}(t, \boldsymbol{x}(t), \boldsymbol{u}(t)), \quad t \in \tilde{I}, \tag{4.28}$$

with initial condition

$$\boldsymbol{x}(t_0) = \boldsymbol{x}_0, \tag{4.29}$$

where $t$ denotes time, $\boldsymbol{x}(t) \in \mathbb{R}^n$ is the *state* of the system at time $t$, $\boldsymbol{u}(t) \in \mathbb{R}^m$ is the control signal at time $t$, $\boldsymbol{x}_0 \in \mathbb{R}^n$ is a given *initial state*, and $\boldsymbol{f}$ is a given function. The interval $\tilde{I}$ is called the *time horizon* for the system.

The control signal in (4.28)–(4.29) is an input variable to be chosen optimally. The system evolves under the influence of the control signal according to equation (4.28), which expresses the rate of change of the system's state as a function of the current time, the current state, and the current value of the control signal. Thus, the control signal drives the system's evolution from $t = t_0$ to $t = t_f$.

In any practical system, the control signal cannot be unbounded. Thus, the bound constraints are typically imposed on the control signal

$$\boldsymbol{u}_{\min} \leqslant \boldsymbol{u}(t) \leqslant \boldsymbol{u}_{\max}, \quad t \in \tilde{I}. \tag{4.30}$$

Any Borel measurable function $\boldsymbol{u} : \tilde{I} \to \mathbb{R}^m$ satisfying (4.30) is called an *admissible control*. Let $\mathcal{U}$ denote the class of all such admissible controls.

The physical constraints can be modeled by the following *canonical constraints*

$$g_j(\boldsymbol{u}) = \Phi_j(\boldsymbol{x}(t_f)) + \int_{t_0}^{t_f} \mathcal{L}_j(t, \boldsymbol{x}(t), \boldsymbol{u}(t))\mathrm{d}t = 0 \text{ or } \geqslant 0, \quad j \in \mathcal{C}, \tag{4.31}$$

where $\mathcal{C}$ is a finite index set and $\Phi_j$ and $\mathcal{L}_j$ are given functions.

The canonical form (4.31) encapsulates many of the common constraints arising in practice. For example, *terminal state constraints* of the form $\boldsymbol{x}(t_f) = \boldsymbol{x}_f$, where $\boldsymbol{x}_f$ is a desired terminal state, can be modeled by a canonical equality constraint with $\Phi_j = \|\boldsymbol{x}(t_f) - \boldsymbol{x}_f\|^2$ and $\mathcal{L}_j = 0$. Furthermore, *continuous state constraints* of the form $h_j(t, \boldsymbol{x}(t)) \geqslant 0$ can (in theory) be modeled by a canonical equality constraint with $\Phi_j = 0$ and $\mathcal{L}_j = \min^2\{h_j(t, \boldsymbol{x}(t)), 0\}$.

Any admissible control $\boldsymbol{u} \in \mathcal{U}$ satisfying the canonical constraints (4.31) is called a *feasible control*. Let $\mathcal{F}$ denote the class of all such feasible controls.

In order to evaluate the performance of a system quantitatively, the designer selects a performance index. It is often formulated as

$$J(\boldsymbol{u}) = \Phi(t_f, \boldsymbol{x}(t_f)) + \int_{t_0}^{t_f} \mathcal{L}(t, \boldsymbol{x}(t), \boldsymbol{u}(t))\mathrm{d}t, \tag{4.32}$$

and is said to be in the *Bolza form*. If $\Phi = 0$, then it is said to be in the *Lagrange form*, and if $\mathcal{L} = 0$, then in the *Mayer form*.

Now, the standard optimal control problem can be stated as follows:

**Problem 4.1.** Choose a feasible control $\boldsymbol{u} \in \mathcal{F}$ to minimize the performance index (4.32).

### *4.3.2 Optimal Multiprocess Control Problem*

In contrast to the standard optimal control problem, optimal multiprocess control problem is a dynamic optimization problem involving a collection of control systems, coupled through constraints in the endpoints of the constituent state trajectories and through the cost function [52]. In optimal multiprocess control problems, frequent reference is made to points in product spaces and to products of product spaces. In this connection, a point $\left(\left(a^1, b^1, \cdots\right), \left(a^2, b^2, \cdots\right), \cdots, \left(a^k, b^k, \cdots\right)\right)$ is denoted by $\left\{a^i, b^i, \cdots\right\}_{i=1}^k$ or, briefly, $\left\{a^i, b^i, \cdots\right\}$.

Consider a multiprocess described by the ordinary differential equations

$$\dot{\boldsymbol{x}}^i(t) = \boldsymbol{f}^i\left(t, \boldsymbol{x}^i(t), \boldsymbol{w}^i(t)\right), \quad a.e.\ t \in \left[\tau_0^i, \tau_1^i\right], \quad i = 1, 2, \ldots, k, \tag{4.33}$$

where $t$ denotes time, $\boldsymbol{x}^i(t) \in \mathbb{R}^{n_i}$ is the *state* at time $t$, $\boldsymbol{w}^i(t) \in \mathbb{R}^{m_i}$ is the *control signal* at time $t$, and $\boldsymbol{f}^i : \mathbb{R} \times \mathbb{R}^{n_i} \times \mathbb{R}^{m_i} \to \mathbb{R}^{n_i}$ is a given function. We call $\left\{\tau_0^i, \tau_1^i, \boldsymbol{x}^i(\cdot), \boldsymbol{w}^i(\cdot)\right\}$ a point of multiprocess (4.33) if $\tau_0^i$, $\tau_1^i$ are the left and right endpoints of a closed subinterval of $\mathbb{R}$, $\boldsymbol{x}^i(\cdot) : \left[\tau_0^i, \tau_1^i\right] \to \mathbb{R}^{n_i}$ is an absolutely continuous function, and $\boldsymbol{w}^i(\cdot) : \left[\tau_0^i, \tau_1^i\right] \to \mathbb{R}^{m_i}$ is a measurable function.

Let $U^i \subset \mathbb{R} \times \mathbb{R}^{m_i}$ and $X^i \subset \mathbb{R} \times \mathbb{R}^{n_i}$, $i = 1, 2, \ldots, k$. Then

$$\boldsymbol{w}^i(t) \in U_t^i, \quad a.e.\ t \in \left[\tau_0^i, \tau_1^i\right], \ i = 1, 2, \ldots, k \tag{4.34}$$

are the *control constraints*, where $U_t^i$ is the set $\{\boldsymbol{u} |\ (t, \boldsymbol{u}) \in U^i\}$; and

$$\boldsymbol{x}^i(t) \in X_t^i, \quad \text{for all } t \in \left[\tau_0^i, \tau_1^i\right], \ i = 1, 2, \ldots, k \tag{4.35}$$

are the *state constraints*, where $X_t^i$ is the set $\left\{\boldsymbol{x} |\ (t, \boldsymbol{x}) \in X^i\right\}$.

Let

$$\mathcal{L}^i : \mathbb{R} \times \mathbb{R} \times \mathbb{R}^{n_i} \times \mathbb{R}^{n_i} \to \mathbb{R}, \quad i = 1, 2, \ldots, k,$$
$$f : E \to \mathbb{R}$$

be given functions, where $E = \prod_i (\mathbb{R} \times \mathbb{R} \times \mathbb{R}^{n_i} \times \mathbb{R}^{n_i})$, and let

$$\Lambda \subset \prod_i \left\{\{\tau_0^i, \tau_1^i, \boldsymbol{x}_0^i, \boldsymbol{x}_1^i\} \in E |\ \tau_0^i \leqslant \tau_1^i, i = 1, 2, \ldots, k\right\}$$

be a given close set.

Define $\tilde{\boldsymbol{f}}^i = \left[\boldsymbol{f}^i, \mathcal{L}^i\right]$, $i = 1, 2, \ldots, k$. We assume for the multiprocess (4.33)–(4.35) that the following conditions are satisfied:

**Assumption 4.1.** *For each $\boldsymbol{x} \in \mathbb{R}^{n_i}$, $\tilde{\boldsymbol{f}}^i(\cdot, \boldsymbol{x}, \cdot)$ is $(\mathscr{L} \times \mathscr{B})$-measurable for $i = 1, 2, \ldots, k$. Here $\mathscr{L}$ denotes the Lebesgue subsets in $\mathbb{R}$, and $\mathscr{B}$, the Borel subsets in $\mathbb{R}^{m_i}$.*

**Assumption 4.2.** *$U^i$ is a Borel measurable set for $i = 1, 2, \ldots, k$.*

There exists a constant $K > 0$ with the following properties:

**Assumption 4.3.** $\left\|\tilde{f}^i(t,x,w)\right\| \leqslant K$ *whenever* $(t,x,w) \in \mathbb{R} \times X_t^i \times U_t^i$.

**Assumption 4.4.** $\left\|\tilde{f}^i(t,x,w) - \tilde{f}^i(t,x',w)\right\| \leqslant K\|x - x'\|$ *whenever* $(t,x,w)$, $(t,x',w) \in \mathbb{R} \times X_t^i \times U_t^i$.

**Assumption 4.5.** $f$ *is locally Lipschitz continuous.*

We now pose the optimal multiprocess control problem:

**Problem 4.2.** Minimize

$$f\left(\left\{\tau_0^i, \tau_1^i, x^i\left(\tau_0^i\right), x^i\left(\tau_1^i\right)\right\}\right) + \sum_i \int_{\tau_0^i}^{\tau_1^i} \mathcal{L}^i\left(t, x^i(t), w^i(t)\right) dt \tag{4.36}$$

over multiprocesses $\left\{\tau_0^i, \tau_1^i, x^i\left(\tau_0^i\right), x^i\left(\tau_1^i\right)\right\}$ that satisfy $\left\{\tau_0^i, \tau_1^i, x^i\left(\tau_0^i\right), x^i\left(\tau_1^i\right)\right\} \in \Lambda$ and constraints (4.34) and (4.35).

## 4.4 Necessary Optimality Conditions

It is very important for finding an optimal control to establish the necessary conditions for optimality. In this section, we will present some basic results of necessary conditions for optimal control problems involving ordinary differential systems. The interested reader can turn to [54,55,88,119,120,199,238] for necessary conditions involving systems described by delay-differential equations and switched systems.

### *4.4.1 Necessary Conditions for Standard Optimal Control Problem*

For standard optimal control problem involving dynamical system (4.28)–(4.29), some fundamental results of the necessary conditions for optimality owe to Pontryagin and his associates [202]. By introducing the so-called costate $\boldsymbol{\lambda}(t) \in \mathbb{R}^n$, we define the *Hamiltonian function* as

$$H(t,x,u,\boldsymbol{\lambda}) = \mathcal{L}(t,x,u) + \boldsymbol{\lambda}^\top(t) f(t,x,u). \tag{4.37}$$

Let $U$ be a subset of $\mathbb{R}^m$ such that the control $u$ is constrained to lie in $U$, where $U$ is known as the *control restraint set* which is, in general, a compact subset of $\mathbb{R}^m$.

Consider the following optimal control problem:

**Problem 4.3.** Choose a control $\boldsymbol{u} \in \mathcal{U}$ such that the performance index (4.32) is to be minimized and subject to the dynamical system (4.28)–(4.29).

**Theorem 4.10.** *Consider Problem 4.3. If $\boldsymbol{u}^* \in \mathcal{U}$ is an optimal control and $\boldsymbol{x}^*(t)$ and $\boldsymbol{\lambda}^*(t)$ are the corresponding state and costate, then it is necessary that*

(a)

$$\begin{cases} \dot{\boldsymbol{x}}^*(t) = \boldsymbol{f}(t, \boldsymbol{x}^*(t), \boldsymbol{u}^*(t)), \\ \boldsymbol{x}^*(t_0) = \boldsymbol{x}_0; \end{cases} \tag{4.38}$$

(b)

$$\begin{cases} \dot{\boldsymbol{\lambda}}^*(t) = -\left(\dfrac{\partial H\left(t, \boldsymbol{x}^*(t), \boldsymbol{u}^*(t), \boldsymbol{\lambda}^*(t)\right)}{\partial \boldsymbol{x}}\right)^{\top}, \\ \boldsymbol{\lambda}^*(t_f) = \left(\dfrac{\partial \Phi\left(t_f, \boldsymbol{x}^*(t_f)\right)}{\partial \boldsymbol{x}}\right)^{\top}; \end{cases} \tag{4.39}$$

(c)

$$\min_{\boldsymbol{v} \in U} H\left(t, \boldsymbol{x}^*(t), \boldsymbol{v}, \boldsymbol{\lambda}^*(t)\right) = H\left(t, \boldsymbol{x}^*(t), \boldsymbol{u}^*(t), \boldsymbol{\lambda}^*(t)\right) \tag{4.40}$$

*for all $t \in \tilde{I}$, except possible on a finite subset of $\tilde{I}$.*

*Remark 4.3.* Note that the condition (c) in the above theorem reduces to the stationary condition

$$\frac{\partial H\left(t, \boldsymbol{x}^*(t), \boldsymbol{u}^*(t), \boldsymbol{\lambda}^*(t)\right)}{\partial \boldsymbol{u}} = \boldsymbol{0} \tag{4.41}$$

if the Hamiltonian function $H$ is continuously differentiable and $U = \mathbb{R}^m$.

To proceed to the second optimal control theorem, we add an additional terminal constraint to the dynamical system (4.28)–(4.29) as follows

$$\boldsymbol{x}(t_f) = \boldsymbol{x}_f, \tag{4.42}$$

where $\boldsymbol{x}_f$ is a given vector in $\mathbb{R}^n$.

The second problem may now be stated as:

**Problem 4.4.** Subject to the dynamical system (4.28)–(4.29) together with the terminal condition (4.42), find a control $\boldsymbol{u} \in \mathcal{U}$ such that the performance index (4.32) is minimized.

**Theorem 4.11.** *Consider Problem 4.4. If $\boldsymbol{u}^* \in \mathcal{U}$ is an optimal control and $\boldsymbol{x}^*(t)$ and $\boldsymbol{\lambda}^*(t)$ are the corresponding state and costate, then it is necessary that*

(a)

$$\begin{cases} \dot{\boldsymbol{x}}^*(t) = \boldsymbol{f}\,(t, \boldsymbol{x}^*(t), \boldsymbol{u}^*(t)), \\ \boldsymbol{x}^*(t_0) = \boldsymbol{x}_0, \\ \boldsymbol{x}^*(t_f) = \boldsymbol{x}_f; \end{cases} \tag{4.43}$$

(b)

$$\dot{\boldsymbol{\lambda}}^*(t) = -\left(\frac{\partial H\left(t, \boldsymbol{x}^*(t), \boldsymbol{u}^*(t), \boldsymbol{\lambda}^*(t)\right)}{\partial \boldsymbol{x}}\right)^{\top}; \tag{4.44}$$

(c)

$$\min_{\boldsymbol{v} \in U} H\left(t, \boldsymbol{x}^*(t), \boldsymbol{v}, \boldsymbol{\lambda}^*(t)\right) = H\left(t, \boldsymbol{x}^*(t), \boldsymbol{u}^*(t), \boldsymbol{\lambda}^*(t)\right) \tag{4.45}$$

*for all $t \in \tilde{I}$, except possible on a finite subset of $\tilde{I}$.*

*Remark 4.4.* Note that the condition (c) in the Theorems 4.10 and 4.11 may also be written as

$$H\left(t, \boldsymbol{x}^*(t), \boldsymbol{u}^*(t), \boldsymbol{\lambda}^*(t)\right) \leqslant H\left(t, \boldsymbol{x}^*(t), \boldsymbol{v}, \boldsymbol{\lambda}^*(t)\right) \tag{4.46}$$

for all $\boldsymbol{v} \in U$ and for all $t \in \tilde{I}$, except possible on a finite subset of $\tilde{I}$.

### *4.4.2 Necessary Conditions for Optimal Multiprocesses*

For optimal multiprocess control problem involving dynamical system (4.33), a theory of necessary conditions was presented in [52, 53]. As a preliminary step toward presenting the necessary conditions, we introduce generalized gradients and normal cones. These are understood in the sense of Clarke [51].

**Definition 4.14.** Let $N$ be an open subset of $\mathbb{R}^k$, let $\boldsymbol{x}$ be a point in $N$, and let $f : N \to \mathbb{R}$ be a locally Lipschitz continuous function. Then the *generalized gradient* $\partial f(\boldsymbol{x})$ of $f$ at $\boldsymbol{x}$ is the set

$$\partial f(\boldsymbol{x}) = \overline{\text{co}}\left\{ \lim_i \nabla f(\boldsymbol{x}_i) |\ \boldsymbol{x}_i \to \boldsymbol{x}_i \to \boldsymbol{x}, \nabla f(\boldsymbol{x}_i) \text{ exists for } i = 1, 2, \ldots \right\}, \tag{4.47}$$

where, for a set $E$, $\overline{\text{co}}E$ is the closed convex hull of $E$.

Given a closed set $S \subset \mathbb{R}$, $d_C : \mathbb{R}^k \to \mathbb{R}$ denotes the Euclidean distance function

$$d_C(\boldsymbol{x}) = \min_{y \in C} \| \boldsymbol{y} - \boldsymbol{x} \|. \tag{4.48}$$

**Definition 4.15.** Let $C \subset \mathbb{R}^k$ be a closed subset of $\mathbb{R}^k$ and $\boldsymbol{x}$ a point in $C$. Then the *normal cone* to $C$ at $\boldsymbol{x}$, written $N_C(\boldsymbol{x})$, is

$$N_C(\boldsymbol{x}) = \mathrm{cl}\left\{ \bigcup_{\lambda \geq 0} \lambda \partial d_C(\boldsymbol{x}) \right\}, \tag{4.49}$$

where, for a set $E$, cl$E$ is the closure of $E$.

Application of the theory of optimal multiprocesses usually involves analysis of generalized gradients and normal cones. The following identities and estimates will be useful in this regard.

**Proposition 4.1.** (i) *For any closed subset $C \subset \mathbb{R}^k \times \mathbb{R}^l$ and point $(\boldsymbol{a}, \boldsymbol{b}) \in C$, we have*

$$N_{\{(x,x,y)|(x,y)\in C\}}((\boldsymbol{a}, \boldsymbol{a}, \boldsymbol{b})) \subset \{(\boldsymbol{p}, \boldsymbol{q}, \boldsymbol{r})|\ (\boldsymbol{p}+\boldsymbol{q}, \boldsymbol{r}) \in N_C((\boldsymbol{a}, \boldsymbol{b}))\}. \tag{4.50}$$

(ii) *For any closed subset $C \subset \mathbb{R}^k \times \mathbb{R}^l$ and point$(\boldsymbol{a}, \boldsymbol{b}, \boldsymbol{c})$ such that $(\boldsymbol{a} - \boldsymbol{b}, \boldsymbol{c}) \in C$, we have*

$$N_{\{(x,y,z)|(x-y,z)\in C\}}((\boldsymbol{a}, \boldsymbol{b}, \boldsymbol{c})) \subset \{(\boldsymbol{p}, -\boldsymbol{p}, \boldsymbol{r})|\ (\boldsymbol{p}, \boldsymbol{r}) \in N_C((\boldsymbol{a}, \boldsymbol{b}))\}. \tag{4.51}$$

(iii) *For closed sets $C_1 \subset \mathbb{R}^k$ and $C_2 \times \mathbb{R}^l$ and points $\boldsymbol{x} \in C_1$ and $\boldsymbol{y} \in C_2$, we have*

$$N_{C_1 \times C_2}(\boldsymbol{x}, \boldsymbol{y}) = N_{C_1}(\boldsymbol{x}) \times N_{C_2}(\boldsymbol{y}). \tag{4.52}$$

(iv) *Let $\tilde{f} : \mathbb{R}^k \times \mathbb{R}^l \to \mathbb{R}$ be a given locally Lipschitz continuous function. Define $f : \mathbb{R}^k \times \mathbb{R}^k \times \mathbb{R}^l \to \mathbb{R}$ to be*

$$f(\boldsymbol{x}, \boldsymbol{y}, z) = \tilde{f}(\boldsymbol{x} - \boldsymbol{y}, z). \tag{4.53}$$

*Then*

$$\partial f(\boldsymbol{a}, \boldsymbol{b}, \boldsymbol{c}) = \left\{(\boldsymbol{p}, -\boldsymbol{p}, \boldsymbol{q})|\ (\boldsymbol{p}, \boldsymbol{q}) \in \partial \tilde{f}((\boldsymbol{a} - \boldsymbol{b}, \boldsymbol{c}))\right\}. \tag{4.54}$$

**Definition 4.16.** Let $I \subset \mathbb{R}$ be an open interval and let $\boldsymbol{g} : I \to \mathbb{R}^k$ be a measurable function. Take a point $t \in I$. Then the *set of essential values* of $\boldsymbol{g}$ at $t$, denoted $\operatorname{ess}_{s \to t} \boldsymbol{g}(s)$, comprises the points $\boldsymbol{x} \in \mathbb{R}^k$ such that, for any $\epsilon > 0$, the set

$$\{s|\ \|\boldsymbol{x}(s) - \boldsymbol{x}\| < \epsilon\} \tag{4.55}$$

has a positive measure. If a point lies in $\underset{s \to t}{\text{co ess}}\, \boldsymbol{g}(s)$, we say it is a *convex essential value* of $\boldsymbol{g}$ at $t$.

It is clearly the case that if $\boldsymbol{g}$ is continuous at $t$, then

$$\underset{s \to t}{\text{ess}}\, \boldsymbol{g}(s) = \{\boldsymbol{g}(t)\},$$

i.e., the essential value is merely the value of the function.

**Definition 4.17.** Given a set $D \subset \mathbb{R}^l$, a multifunction $\Gamma : D \rightrightarrows \mathbb{R}^k$ is a mapping from $D$ to the subsets of $\mathbb{R}^k$. Its *graph* is the set

$$\text{graph}\Gamma = \left\{(\boldsymbol{x}, \boldsymbol{y}) \mid \boldsymbol{x} \in \mathbb{R}^l, \boldsymbol{y} \in \Gamma(\boldsymbol{x})\right\}. \tag{4.56}$$

Define the Hamiltonian function $H^i$ to be

$$H^i(t, \boldsymbol{x}, \boldsymbol{u}, \boldsymbol{p}, \lambda) = \boldsymbol{p} \cdot \boldsymbol{f}^i(t, \boldsymbol{x}, \boldsymbol{u}) - \lambda \mathcal{L}^i(t, \boldsymbol{x}, \boldsymbol{u}), \quad i = 1, 2, \ldots, k. \tag{4.57}$$

The following maximum principle for solutions to Problem 4.2 can be stated.

**Theorem 4.12.** *Let* $\left\{T_0^i, T_1^i, \boldsymbol{x}^i(\cdot), \boldsymbol{u}^i(\cdot)\right\}$ *be a solution to Problem* 4.2. *Suppose that*

$$\text{graph}\left\{\boldsymbol{x}^i(\cdot)\right\} \subset \text{interior}\left\{X^i\right\}$$

*for* $i = 1, 2, \ldots, k$, *where, for a set* $E$, interior$E$ *is the interior of* $E$, *and that Assumptions* 4.1–4.5 *are satisfied. Then there exist a real number* $\lambda$ *equal to zero or one, real numbers* $h_0^i, h_1^i$, $i = 1, 2, \ldots, k$, *and absolutely continuous functions* $\boldsymbol{p}^i(\cdot) : \left[T_0^i, T_1^i\right] \to \mathbb{R}^{n_i}$ *for* $i = 1, 2, \ldots, k$, *such that* $\lambda + \sum_i \left\|\boldsymbol{p}^i\left(T_1^i\right)\right\| > 0$,

$$-\dot{\boldsymbol{p}}^i(t) \in \partial_x H^i\left(t, \boldsymbol{x}^i(t), \boldsymbol{u}^i(t), \boldsymbol{p}^i(t), \lambda\right), \quad a.e.\ t \in \left[T_0^i, T_1^i\right], \tag{4.58}$$

$$H^i\left(t, \boldsymbol{x}^i(t), \boldsymbol{u}^i(t), \boldsymbol{p}^i(t), \lambda\right) = \max_{\boldsymbol{w} \in U_t^i} H^i\left(t, \boldsymbol{x}^i(t), \boldsymbol{w}, \boldsymbol{p}^i(t), \lambda\right), \quad a.e.\ t \in \left[T_0^i, T_1^i\right], \tag{4.59}$$

$$h_0^i \in \underset{t \to T_0^i}{\text{co ess}} \left[\sup_{\boldsymbol{w} \in U_t^i} H^i\left(t, \boldsymbol{x}^i\left(T_0^i\right), \boldsymbol{w}, \boldsymbol{p}^i\left(T_0^i\right), \lambda\right)\right], \tag{4.60}$$

$$h_1^i \in \underset{t \to T_1^i}{\text{co ess}} \left[\sup_{\boldsymbol{w} \in U_t^i} H^i\left(t, \boldsymbol{x}^i\left(T_1^i\right), \boldsymbol{w}, \boldsymbol{p}^i\left(T_1^i\right), \lambda\right)\right], \tag{4.61}$$

*for* $i = 1, 2, \ldots, k$, *and*

$$\left\{-h_0^i, h_1^i, \boldsymbol{p}^i\left(T_0^i\right), -\boldsymbol{p}^i\left(T_1^i\right)\right\} \in N_\Lambda + \lambda \partial f \tag{4.62}$$

*where* $\partial_x H^i$ *denotes the partial generalized gradient in the second variable and the normal cone* $N_\Lambda$ *and the generalized gradient* $\partial f$ *are evaluated at* $\left\{T_0^i, T_1^i, \boldsymbol{x}^i\left(T_0^i\right), \boldsymbol{x}^i\left(T_1^i\right)\right\}$.

The Problem 4.2 is *autonomous* when the functions $\boldsymbol{f}^i$ and $\mathcal{L}^i$ have no dependence on $t$ and when the control set $U_t^i$ is the same set $U_0^i$ for all $t$.

**Corollary 4.1.** *Under the assumptions of Theorem* 4.12, *when in addition Problem* 4.2 *is autonomous, then the conclusions of the theorem can be supplemented by the following. For each* $i$, *there is a constant* $h^i$ *such that* $h_0^i = h_1^i = h^i$ *and*

$$h^i = \sup_{w \in U_0^i} H^i\left(\boldsymbol{x}^i(t), \boldsymbol{w}, \boldsymbol{p}^i(t), \lambda\right), \quad \textit{for all } t \in \left[T_0^i, T_1^i\right]. \tag{4.63}$$

# Chapter 5
# Optimal Control of Nonlinear Multistage Systems

## 5.1 Introduction

In this chapter, we consider the multistage optimal control problem in fed-batch fermentation. As a case study, the microbial conversion of glycerol to 1,3-propanediol (1,3-PD) in fed-batch culture is investigated. This process is particularly attractive in that the process is relatively easy and does not generate toxic by-products. Glycerol can be converted to 1,3-PD by several microorganisms [31, 207]. In the actual fermentation process, the fed-batch culture begins with a batch culture. After the exponential growth phase (i.e., a period in which the number of new bacteria appearing per unit time is proportional to the present population), glycerol and alkali are continuously added to the fermentor. This helps to maintain a suitable environment for cell growth. At the end of the feeding, another batch phase starts again. The above processes are repeated until the end of the final batch phase.

Modeling the fermentation process is prerequisite to carrying out optimal control and to improving the productivity of the product. In this chapter, taking the feeding of glycerol as a time-continuous process, we propose a controlled nonlinear multistage system to describe the microbial fed-batch culture. Compared with the existing systems, this system is much closer to the actual fermentation process. Furthermore, to maximize the concentration of 1,3-PD at the terminal time, we present an optimal control model subject to the proposed controlled multistage system and continuous state inequality constraints. The existence of optimal control is ascertained using the theory of bounded variation. By the way, there exist many methods to solve the problem of optimal feeding rate, such as Luus–Jaakola search method [163], multiple shooting technique [191], genetic algorithm [217], and so on. However, these methods are all applied to the fed-batch process with one single operation in which the substrates are fed to the fermentor continuously. In some fermentation processes, especially in glycerol bioconversion to 1,3-PD



C. Liu, Z. Gong, *Optimal Control of Switched Systems Arising in Fermentation Processes*, Springer Optimization and Its Applications 97,
DOI 10.1007/978-3-662-43793-3_5

fermentation process, substrates are intermittently fed into the fermentor leading to serial fed-batch operations. As a result, the computation is more complex and it is necessary to present a new method to solve this class of problems.

In this chapter, applying the extended control parameterization method, we obtain a sequence of approximate parameter optimization problems. The convergence analysis of this approximation is also investigated. Based on the above discretization and an improved particle swarm optimization (PSO) algorithm, a global optimization algorithm is constructed to solve the optimal control model. Numerical results show that the concentration of 1,3-PD at the terminal time can be increased considerably compared with the experimental data.

The main references of this chapter are [143] and [150].

## 5.2 Controlled Multistage Systems

In the fed-batch process, the composition of the culture medium, cultivation conditions, and analytical methods of fermentative products were similar to those previously reported [48]. According to the fermentation process, we assume that

**Assumption 5.1.** *The concentrations of reactants are uniform in the reactor. Time delay and nonuniform space distribution are ignored.*

**Assumption 5.2.** *During the process of fed-batch culture, only glycerol and alkali are fed into the reactor. Moreover, the feeding velocity ratio of alkali to glycerol is a constant.*

**Assumption 5.3.** *The feeding rates of glycerol and alkali are bounded and have bounded variations on each time interval of feeding process.*

Let $\boldsymbol{x}(t) := (x_1(t), x_2(t), x_3(t), x_4(t), x_5(t))^{\top} \in \mathbb{R}^5_+, t \in [0, T]$, be the state vector. The components of $\boldsymbol{x}(t)$ represent the extracellular concentrations of biomass, glycerol, 1,3-PD, acetic acid, and ethanol at time $t$ in the fermentor, respectively. $T$ is the terminal time of the fermentation and $\boldsymbol{x}_0$ is a given initial state. Since glycerol and alkali are fed to the fermentor at a proportional constant $r$, the feeding rate of alkali can be determined by that of glycerol. Let $u(t)$, the feeding rate of glycerol, be the control function. Let $t_i, i \in \Lambda := \{1, 2, \dots, 2N + 1\}$, be the switching instants such that $0 = t_0, t_{i-1} < t_i, i \in \Lambda$, and $t_{2N+1} = T$, which is decided a priori in the experiment. In particular, $t_{2j+1}$ is the moment of adding glycerol, at which the fermentation process switches to the feeding process from the batch process, and $t_{2j+2}$ denotes the moment of ending the flow of glycerol, at which the fermentation process jumps into the batch process from the feeding process, $j \in \bar{\Lambda}_1 := \{0, 1, 2, \dots, N - 1\}$. Under the Assumptions 5.1–5.3, mass balances of biomass, substrate, and products in fed-batch culture can be formulated as the following controlled multistage system

$$\begin{cases} \dot{\boldsymbol{x}}(t) = \boldsymbol{f}^i(t, \boldsymbol{x}(t), u(t)), \\ \boldsymbol{x}(t_{i-1}+) = \boldsymbol{x}(t_{i-1}),\ t \in (t_{i-1}, t_i],\ i = 1, 2, \ldots, 2N+1, \\ \boldsymbol{x}(0) = \boldsymbol{x}_0, \end{cases} \tag{5.1}$$

where the notation + indicates the limit from the right, and for $t \in (t_{2j}, t_{2j+1}]$, $j \in \bar{\Lambda}_2 := \{0, 1, \ldots, N\}$,

$$f_\ell^{2j+1}(t, \boldsymbol{x}(t), u(t)) = q_\ell(\boldsymbol{x}(t))x_1(t), \quad \ell = 1, 3, 4, 5, \tag{5.2}$$

and

$$f_2^{2j+1}(t, \boldsymbol{x}(t), u(t)) = -q_2(\boldsymbol{x}(t))x_1(t); \tag{5.3}$$

for $t \in (t_{2j+1}, t_{2j+2}]$, $j \in \bar{\Lambda}_1$,

$$f_\ell^{2j+2}(t, \boldsymbol{x}(t), u(t)) = q_\ell(\boldsymbol{x}(t))x_1(t) - D\big(t, u(t)\big)x_\ell(t), \quad \ell = 1, 3, 4, 5, \tag{5.4}$$

and

$$f_2^{2j+2}(t, \boldsymbol{x}(t), u(t)) = D\big(t, u(t)\big)\left(\frac{c_{s0}}{r+1} - x_2(t)\right) - q_2(\boldsymbol{x}(t))x_1(t). \tag{5.5}$$

In (5.4) and (5.5), $D(t, u(t))$ is the dilution rate at time $t$. $r$ is the velocity ratio of adding alkali to glycerol. $c_{s0}$ denotes the initial concentration of glycerol in feed. Furthermore,

$$D(t, u(t)) = \frac{(1+r)u(t)}{V(t)}, \tag{5.6}$$

$$V(t) = V_0 + \int_0^t (1+r)u(s)\mathrm{d}s. \tag{5.7}$$

In (5.6)–(5.7), $V_0$ is the initial volume of solution in the fermentor. On the basis of the previous work [271], the specific growth rate of cells $q_1(\boldsymbol{x}(t))$ is expressed by

$$q_1(\boldsymbol{x}(t)) = \frac{\Delta_1 x_2(t)}{x_2(t) + k_1} \prod_{\ell=2}^{5} \left(1 - \frac{x_\ell(t)}{x_\ell^*}\right)^{n_\ell}, \tag{5.8}$$

where $\Delta_1$ is the maximum specific growth rate, $k_1$ is the Monod saturation constant, $x_\ell^*$ are the maximal residual substrate or product concentrations, and $n_\ell$ are the exponents for the substrate or products. The specific consumption rate of substrate $q_2(\boldsymbol{x}(t))$ is

$$q_2(\boldsymbol{x}(t)) = m_2 + q_1(\boldsymbol{x}(t))Y_2 + \frac{\Delta_2 x_2(t)}{x_2(t) + k_2}. \tag{5.9}$$

In (5.9), $m_2$ is the maintenance term of substrate consumption under substrate-limited conditions. $Y_2$ is the maximum growth yield. $\Delta_2$ is the maximum increment of substrate consumption rate under substrate-sufficient conditions. $k_2$ is the saturation constant for substrate. The specific formation rates of 1,3-PD and acetate $q_\ell(\boldsymbol{x}(t)), \ell = 3, 4$, are defined as

$$q_\ell(\boldsymbol{x}(t)) = -m_\ell + q_1(\boldsymbol{x}(t))Y_\ell + \frac{\Delta_\ell x_2(t)}{x_2(t) + k_\ell}, \tag{5.10}$$

where $m_\ell$ are the maintenance terms of product formations under substrate-limited conditions, $Y_\ell$ are the maximum product yields, $\Delta_\ell$ are the maximum increments of product formation rates under substrate-sufficient conditions, and $k_\ell$ are saturation constants for products. Moreover, the specific formation rate of ethanol $q_5(\boldsymbol{x}(t))$ can be described by

$$q_5(\boldsymbol{x}(t)) = q_2(\boldsymbol{x}(t))\left(\frac{c_1}{c_2 + q_1(\boldsymbol{x}(t))x_2(t)} + \frac{c_3}{c_4 + q_1(\boldsymbol{x}(t))x_2(t)}\right), \tag{5.11}$$

in which $c_1, c_2, c_3$, and $c_4$ are parameters for the determination of yield of ethanol on glycerol.

Under anaerobic conditions at 37 °C and pH 7.0, the critical concentrations for cell growth and the parameters in (5.8)–(5.11) are listed in Table 5.1.

Define

$$U := \{u(t)|\, a_i \leqslant u(t) \leqslant b_i, t \in (t_{i-1}, t_i], i \in \Lambda\}, \tag{5.12}$$

where $a_{2j+1}$ and $b_{2j+1}, j \in \bar{\Lambda}_2$, are identically equal to zero and $a_{2j+2}$ and $b_{2j+2}, j \in \bar{\Lambda}_1$, are positive constants which denote the minimal and maximal rates of adding glycerol, respectively. Clearly, $U$ is convex and compact. Any function $u$ from $[0, T]$ into $\mathbb{R}$ such that $u(t) \in U$ and $u$ is of bounded variation on $[t_{i-1}, t_i], i \in \Lambda$, is called an admissible control. Let $\mathcal{U}$ be the class of all such admissible controls. Moreover, the concentrations of biomass, glycerol, and products are restricted in a certain range according to the fermentation, so we consider the properties of the system with state in $W := \prod_{\ell=1}^{5} \left[x_{*\ell}, x_\ell^*\right]$.

**Table 5.1** The parameters and critical concentrations in the system (5.1)

| $\ell$ | $m_\ell$ | $Y_\ell$ | $\Delta_\ell$ | $k_\ell$ | $n_\ell$ | $c_\ell$ | $x_{*\ell}$ | $x_\ell^*$ |
|---|---|---|---|---|---|---|---|---|
| 1 | — | — | 0.67 | 0.28 | — | 0.025 | 0.01 | 6 |
| 2 | 2.20 | 113.6 | 28.58 | 11.43 | 1 | 0.06 | 15 | 2,039 |
| 3 | 2.69 | 67.69 | 26.59 | 15.50 | 3 | 5.18 | 0 | 1,036 |
| 4 | 0.97 | 33.07 | 5.74 | 85.71 | 3 | 50.45 | 0 | 1,026 |
| 5 | — | — | — | — | 3 | — | 0 | 360.9 |

## 5.3 Properties of the Controlled Multistage Systems

In this section, we will prove some important properties of the solution to the system (5.1), such as existence, continuity with respect to control function, and so on. Let $L_\infty$ denote the Banach space $L_\infty([0,T],\mathbb{R})$ of all essentially bounded functions from $[0,T]$ into $\mathbb{R}$. Its norm is

$$\|u\|_\infty := \operatorname*{ess\,sup}_{t\in[0,T]} \|u(t)\|,$$

where $\|\cdot\|$ is the Euclidean norm.

Firstly, we discuss some properties of the function $\boldsymbol{f}^i, i \in \Lambda$.

**Proposition 5.1.** *The functions* $\boldsymbol{f}^i : [0,T]\times\mathbb{R}^5_+\times U \to \mathbb{R}^5$, $i = 1,2,\ldots,2N+1$, *defined in* (5.2)–(5.5) *satisfy*

(a) $\boldsymbol{f}^i$, *together with their partial derivatives with respect to* $\boldsymbol{x}$ *and* $u$, *are continuous on* $[0,T]$ *for each* $(\boldsymbol{x},u) \in \mathbb{R}^5_+ \times U$ *and continuous on* $\mathbb{R}^5_+ \times U$ *for each* $t \in [0,T]$;
(b) *There exists a positive constant* $K$ *such that*

$$\|\boldsymbol{f}^i(t,\boldsymbol{x},u)\| \leqslant K(1+\|\boldsymbol{x}\|), \forall (t,\boldsymbol{x},u) \in [0,T]\times\mathbb{R}^5_+\times U. \tag{5.13}$$

*Proof.* (a) This conclusion can be obtained by the expressions of $\boldsymbol{f}^i, i = 1,2,\ldots,2N+1$, in (5.2)–(5.5).
(b) For any $\boldsymbol{x}(t) \in \mathbb{R}^5_+$ and $u(t) \in U$, we know that

$$\begin{aligned}|f_1^i(t,\boldsymbol{x}(t),u(t))| &\leqslant |q_1(\boldsymbol{x}(t))|\cdot|x_1(t)| + |D(t,u(t))|\cdot|x_1(t)|\\ &\leqslant \left(\Delta_1 + \frac{(1+r)b}{V_0}\right)\|\boldsymbol{x}(t)\|,\end{aligned}$$

where $b := \max_i\{b_i\}$. Letting $L_1 := |\Delta_1| + \left|\frac{(1+r)b}{V_0}\right|$, we conclude that $|f_1^i(t,\boldsymbol{x}(t),u(t))| \leqslant L_1\|\boldsymbol{x}(t)\|$.

$$\begin{aligned}|f_2^i(t,\boldsymbol{x}(t),u(t))| &\leqslant \left|D(t,u(t))\frac{c_{s0}}{r+1}\right| + |D(t,u(t))|\cdot|x_2(t)|\\ &\quad + |q_2(\boldsymbol{x}(t))|\cdot|x_1(t)|\\ &\leqslant \frac{c_{s0}b}{V_0} + \frac{(1+r)b}{V_0}|x_2(t)|\\ &\quad + (|m_2| + |\Delta_1|\cdot|Y_2| + |\Delta_2|)|x_1(t)|.\end{aligned}$$

Letting $L_2 := \max\left\{\frac{(1+r)b}{V_0}, |m_2| + |\Delta_1|\cdot|Y_2| + |\Delta_2|\right\}$, we conclude that $\left|f_2^i(t, \boldsymbol{x}(t), u(t))\right| \leqslant L_2\|\boldsymbol{x}(t)\| + \frac{bc_{s0}}{V_0}$.
For $\ell = 3, 4$,

$$\begin{aligned}\left|f_\ell^i(t, \boldsymbol{x}(t), u(t))\right| &\leqslant |q_\ell(\boldsymbol{x}(t))|\cdot|x_1(t)| + |D(t, u(t))|\cdot|x_\ell(t)| \\ &\leqslant (|m_\ell| + |\Delta_1|\cdot|Y_\ell| + |\Delta_\ell|)|x_1(t)| + \frac{(1+r)b}{V_0}|x_\ell(t)|.\end{aligned}$$

Letting $L_\ell := \max\left\{|m_\ell| + |\Delta_1|\cdot|Y_\ell| + |\Delta_\ell|, \frac{(1+r)b}{V_0}\right\}$, we conclude that $\left|f_\ell^i(t, \boldsymbol{x}(t), u(t))\right| \leqslant L_\ell\|\boldsymbol{x}(t)\|$.

$$\begin{aligned}\left|f_5^i(t, \boldsymbol{x}(t), u(t))\right| &\leqslant |q_5(\boldsymbol{x}(t))|\cdot|x_1(t)| + |D(t, u(t))|\cdot|x_5(t)| \\ &\leqslant \left(\left|\frac{c_1}{c_2}\right| + \left|\frac{c_3}{c_4}\right|\right)(|m_2| + |\Delta_1|\cdot|Y_2| + |\Delta_2|)|x_1(t)| \\ &\quad + \frac{(1+r)b}{V_0}|x_5(t)|.\end{aligned}$$

Letting $L_5 := \max\left\{\left(\left|\frac{c_1}{c_2}\right| + \left|\frac{c_3}{c_4}\right|\right)(|m_2| + |\Delta_1|\cdot|Y_2| + |\Delta_2|), \frac{(1+r)b}{V_0}\right\}$, we conclude that $|f_5^i(t, \boldsymbol{x}(t), u(t))| \leqslant L_5\|\boldsymbol{x}(t)\|$.
Set $K := \max\left\{L_1, L_2, \ldots, L_5, \frac{bc_{s0}}{V_0}\right\}$. Then we have

$$\|\boldsymbol{f}^i(t, \boldsymbol{x}(t), u(t))\| \leqslant K(1 + \|\boldsymbol{x}(t)\|).$$ □

Now, we can obtain the existence and uniqueness of the solution to the system (5.1).

**Theorem 5.1.** *For each $u \in \mathcal{U}$, the controlled multistage system* (5.1) *has a unique solution denoted by $\boldsymbol{x}(\cdot|u)$. Moreover, $\boldsymbol{x}(\cdot|u)$ satisfies the following integral equation*

$$\boldsymbol{x}(t|u) = \boldsymbol{x}(t_{i-1}|u) + \int_{t_{i-1}}^{t} \boldsymbol{f}^i(s, \boldsymbol{x}(s|u), u(s))\mathrm{d}s, \ \forall t \in (t_{i-1}, t_i], \ i \in \Lambda. \quad (5.14)$$

*Proof.* The proof can be obtained from Proposition 5.1 and the theory of differential equations [5]. □

**Theorem 5.2.** *Given the initial state $\boldsymbol{x}_0$ and for all $u \in \mathcal{U}$, the unique solution $\boldsymbol{x}(\cdot|u)$ of the system* (5.1) *is uniformly bounded on* $[0, T]$ *and Lipschitz continuous in $u$ for all $t \in [0, T]$.*

*Proof.* In view of Theorem 5.1 and Proposition 5.1, we obtain that for each $u \in \mathcal{U}$,

$$\begin{aligned}\|\boldsymbol{x}(t|u)\| &\leqslant \|\boldsymbol{x}_0\| + \sum_{j=1}^{i-1}\int_{t_{j-1}}^{t_j} \|\boldsymbol{f}^j(s, \boldsymbol{x}(s|u), u(s))\| \mathrm{d}s + \int_{t_{i-1}}^{t} \|\boldsymbol{f}^i(s, \boldsymbol{x}(s|u), u(s))\| \mathrm{d}s \\ &\leqslant \|\boldsymbol{x}_0\| + K\int_0^t (1 + \|\boldsymbol{x}(s|u)\|)\mathrm{d}s, \quad \forall t \in [0, T].\end{aligned}$$

By Lemma 4.1, it follows that

$$\|\boldsymbol{x}(t|u)\| \leqslant (\|\boldsymbol{x}_0\| + KT)\exp(KT), \quad \forall t \in [0, T].$$

which gives a value,

$$M' := (\|\boldsymbol{x}_0\| + KT)\exp(KT),$$

for the uniformly bounded property.

Let $v, \tilde{v}$ be two distinct control functions in $\mathcal{U}$. Applying the mean value inequality, we have

$$\|\boldsymbol{f}^i(t, \boldsymbol{x}(t|v), v(t)) - \boldsymbol{f}^i(t, \boldsymbol{x}(t|\tilde{v}), \tilde{v}(t))\| \leqslant \gamma_1\|\boldsymbol{x}(t|v) - \boldsymbol{x}(t|\tilde{v})\| + \gamma_2\|v(t) - \tilde{v}(t)\|,$$

where $\gamma_1$ and $\gamma_2$ are upper bounds for $\|\boldsymbol{f}^i_{\boldsymbol{x}}\|$ and $\|\boldsymbol{f}^i_{\boldsymbol{u}}\|$ on $[0, T] \times \{\boldsymbol{x} \in \mathbb{R}^5_+ | \|\boldsymbol{x}\| \leqslant M'\} \times U$, respectively. Thus,

$$\begin{aligned}\|\boldsymbol{x}(t|v) - \boldsymbol{x}(t|\tilde{v})\| &\leqslant \sum_{j=1}^{i-1}\int_{t_{j-1}}^{t_j} \|\boldsymbol{f}^j(s, \boldsymbol{x}(s|v), v(s)) - \boldsymbol{f}^j(s, \boldsymbol{x}(s|\tilde{v}), \tilde{v}(s))\| \mathrm{d}s \\ &\quad + \int_{t_{i-1}}^{t} \|\boldsymbol{f}^i(s, \boldsymbol{x}(s|v), v(s)) - \boldsymbol{f}^i(s, \boldsymbol{x}(s|\tilde{v}), \tilde{v}(s))\| \mathrm{d}s \\ &\leqslant \int_0^t (\gamma_1\|\boldsymbol{x}(s|v) - \boldsymbol{x}(s|\tilde{v})\| + \gamma_2\|v(s) - \tilde{v}(s)\|)\mathrm{d}s.\end{aligned}$$

By Lemma 4.1, we conclude that, for all $v, \tilde{v} \in \mathcal{U}$,

$$\begin{aligned}\|\boldsymbol{x}(t|v) - \boldsymbol{x}(t|\tilde{v})\| &\leqslant K'\int_0^T \|v(s) - \tilde{v}(s)\| \mathrm{d}s \\ &\leqslant K'T\|v - \tilde{v}\|_\infty, \ \forall t \in [0, T],\end{aligned}$$

where $K' := \gamma_2 \exp(\gamma_1 T)$. This yields the Lipschitz continuity of the solution to the system (5.1). □

## 5.4 Optimal Control Models

For mathematical convenience, we define the set of solutions to the system (5.1), $\mathcal{S}_0$, as follows:

$$\mathcal{S}_0 = \{\boldsymbol{x}(\cdot|u) | \ \boldsymbol{x}(\cdot|u) \text{ is a solution of the system (5.1) for } u \in \mathcal{U}\}. \tag{5.15}$$

Since the concentrations of biomass, glycerol, and products are restricted in $W$ in the fermentation, the set of admissible solutions is

$$\mathcal{S} = \{\boldsymbol{x}(\cdot|u) \in \mathcal{S}_0 | \ \boldsymbol{x}(t|u) \in W \text{ for all } t \in [0, T]\}. \tag{5.16}$$

Let the set of feasible controls corresponding to $\mathcal{S}$ be

$$\mathcal{F} = \{u \in \mathcal{U} | \ \boldsymbol{x}(\cdot|u) \in \mathcal{S}\}. \tag{5.17}$$

In fed-batch culture of glycerol bio-dissimilation to 1,3-PD, the problem, to optimize the rate of infused glycerol such that the concentration of 1,3-PD at the terminal time is as high as possible, can be formulated as follows:

$$\text{(MOCP)} \quad \min \quad J(u) = -x_3(T|u)$$
$$\text{s.t.} \quad u \in \mathcal{F},$$

where $x_3(\cdot|u)$ is the third component of the solution to the system (5.1).

By the theory of bounded variation, we can show the existence of optimal control for (MOCP).

**Lemma 5.1.** *Under Assumption* 5.3, *the following results hold.*

(a) *For any* $u \in \mathcal{U}$, $u \in BV([0, T])$.
(b) *Given any sequence* $\{u_n\}$ *in* $\mathcal{U}$, *there exists a constant* $M > 0$ *such that*

$$|u_n(t)| \leqslant M \textit{ for all } t \in [0, T] \textit{ and } n = 1, 2, \ldots, \tag{5.18}$$

*and*

$$\bigvee_0^T u_n(t) \leqslant M \tag{5.19}$$

*for each* $n$.

*Proof.* (a) Since the control function $u$ is of bounded variation on each time interval of the feeding process and is identically equal to zero in each batch process, we conclude that, for any partition $p = \{\tau_0, \tau_1, \ldots, \tau_{n_p}\}$ of $[0, T]$,

$$\sum_{i=1}^{n_p} |u(\tau_i) - u(\tau_{i-1})| \leqslant \sum_{j=0}^{N-1} \bigvee_{t_{2j+1}}^{t_{2j+2}} u(t) + 3Nb, \ \forall u \in \mathcal{U},$$

where, as in proof of Proposition 5.1, $b := \max_{j \in \Lambda}\{b_j\}$. Furthermore, we have

$$\sum_{i=1}^{n_p} |u(\tau_i) - u(\tau_{i-1})| \leqslant NL + 3Nb, \tag{5.20}$$

where $L := \max_{j \in \{0,1,\ldots,N-1\}} \bigvee_{t_{2j+1}}^{t_{2j+2}} u(t)$. This also indicates that

$$\bigvee_{0}^{T} u(t) < +\infty, \ \forall u \in \mathcal{U}. \tag{5.21}$$

Thus, the conclusion follows from (5.20) and (5.21).

(b) Given any sequence $\{u_n\}$, we obtain

$$|u_n(t)| \leqslant b, \ \forall t \in [0, T] \text{ and } n = 1, 2, \ldots.$$

Let $M := \max\{b, NL + 3Nb\}$. Since the choice of the boundedness of total variation in $(a)$ is independent of $u(t)$, we must conclude that (5.18) and (5.19) hold. □

On this basis, we can give the following existence theorem of optimal control for (MOCP).

**Theorem 5.3.** *Under Assumption* 5.3, (MOCP) *has at least one optimal solution.*

*Proof.* For any sequence $\{u_n\}$ with $u_n \in \mathcal{F} \subseteq \mathcal{U}$, it follows from Lemma 5.1 that $\{u_n\}$ is uniformly bounded with equibounded total variation. By Theorem 2.14, there exists a subsequence, $\{u_{n_k}\}$, of the sequence $\{u_n\}$ that converges to $u^*$ pointwise in $[0, T]$. Clearly, $u^* \in \mathcal{U}$. According to Theorem 5.1, we know there exist solutions $\boldsymbol{x}\left(\cdot|u_{n_k}\right)$ and $\boldsymbol{x}(\cdot|u^*)$ corresponding to $u_{n_k}$ and $u^*$, respectively. Since the solution is continuous in control function $u$ by Theorem 5.2,

$$\lim_{k \to \infty} \boldsymbol{x}\left(\cdot|u_{n_k}\right) = \boldsymbol{x}(\cdot|u^*), \text{ pointwise in } [0, T].$$

Furthermore, since $W$ is compact, $\boldsymbol{x}(t|u^*) \in W$. Hence, $\mathcal{F}$ is a compact set. In view of the continuity of cost function $J(u)$ in $u$, (MOCP) must have at least one optimal solution. □

## 5.5 Computational Approaches

In this section, we will develop a computational approach for solving the (MOCP) by extending the control parameterization technique introduced in [240–242].

For each $p_i \geqslant 1, i \in \Lambda$, let the time subinterval $[t_{i-1}, t_i]$ be partitioned into $n_{p_i}$ subintervals with $n_{p_i} + 1$ partition points denoted by

$$\tau_0^{p_i}, \tau_1^{p_i}, \ldots, \tau_{n_{p_i}}^{p_i}, \tau_0^{p_i} = t_{i-1}, \tau_{n_{p_i}}^{p_i} = t_i, \text{ and } \tau_{k-1}^{p_i} < \tau_k^{p_i}.$$

Let $n_{p_i}$ and $\tau_k^{p_i}$ be chosen such that

(a) $n_{p_i+1} \geqslant n_{p_i}$ ;
(b) $\lim_{p_i \to \infty} |\tau_k^{p_i} - \tau_{k-1}^{p_i}| = 0$.

We now approximate the control function by piecewise constant functions as follows:

$$u^p(t|\sigma^p) = \sum_{i=1}^{2N+1} \sum_{k=1}^{n_{p_i}} \sigma^{p_i,k} \chi_{(\tau_{k-1}^{p_i}, \tau_k^{p_i}]}(t). \tag{5.22}$$

Here, as in Chap. 2, $\chi_{(\tau_{k-1}^{p_i}, \tau_k^{p_i}]}$ denotes the indicator function on the interval $(\tau_{k-1}^{p_i}, \tau_k^{p_i}]$, i.e.,

$$\chi_{(\tau_{k-1}^{p_i}, \tau_k^{p_i}]}(t) = \begin{cases} 1, \ t \in (\tau_{k-1}^{p_i}, \tau_k^{p_i}], \\ 0, \text{ otherwise.} \end{cases}$$

Let

$$\boldsymbol{\sigma}^p := \left((\boldsymbol{\sigma}^{p_1})^\top, (\boldsymbol{\sigma}^{p_2})^\top, \ldots, (\boldsymbol{\sigma}^{p_{2N+1}})^\top\right)^\top,$$

where $\boldsymbol{\sigma}^{p_i} := (\sigma^{p_i,1}, \ldots, \sigma^{p_i,n_{p_i}})^\top$. Furthermore, let $\kappa := \sum_{i=1}^{2N+1} n_{p_i}$ and $\mathcal{V}^p$ be the set of all those $u^p(\cdot|\boldsymbol{\sigma}^p)$ expressed by (5.22) with $\boldsymbol{\sigma}^p \in R^\kappa$. Restricting the control in $\mathcal{V}^p$, the control bounds defined in (5.12) become

$$a_i \leqslant \sigma^{p_i,k} \leqslant b_i, k = 1, \ldots, n_{p_i}, i = 1, \ldots, 2N+1. \tag{5.23}$$

Let $\Xi^p$ be the set of all those $\boldsymbol{\sigma}^p$ vectors which satisfy the constraints (5.23) and $\mathcal{U}^p$ be the set of all those $u^p(\cdot|\boldsymbol{\sigma}^p) \in \mathcal{V}^p$ with $\boldsymbol{\sigma}^p \in \Xi^p$.

With $u \in \mathcal{U}^p$, the controlled multistage system (5.1) takes the form

$$\begin{cases} \dot{\boldsymbol{x}}(t) = \tilde{\boldsymbol{f}}(t, \boldsymbol{x}(t), \boldsymbol{\sigma}^p), \\ \boldsymbol{x}(t_{i-1}+) = \boldsymbol{x}(t_{i-1}), \ t \in (t_{i-1}, t_i], \ i = 1, 2, \ldots, 2N+1, \\ \boldsymbol{x}(0) = \boldsymbol{x}_0, \end{cases} \tag{5.24}$$

where

$$\tilde{f}(t, x(t), \sigma^p) := \sum_{i=1}^{2N+1} f^i\left(t, x(t), \sum_{k=1}^{n_{p_i}} \sigma^{p_i,k}\chi_{(\tau_{k-1}^{p_i}, \tau_k^{p_i}]}(t)\right)\chi_{(t_{i-1}, t_i]}(t).$$

Let $x(\cdot|\sigma^p)$ be the solution of the system (5.24) corresponding to the control parameter vector $\sigma^p \in \Xi^p$. Similarly, the constraint in (5.16) can be rewritten as

$$x(t|\sigma^p) \in W. \tag{5.25}$$

The set of feasible states and the set of feasible controls used previously become

$$\mathcal{S}^p = \{x(\cdot|\sigma^p)|\ x(\cdot|\sigma^p) \text{ is a solution of (5.24) and } x(t|\sigma^p) \in W \text{ for all } t \in [0, T]\},$$
$$\mathcal{F}^p = \{\sigma^p \in \Xi^p|\ x(\cdot|\sigma^p) \in \mathcal{S}^p\}.$$

Now, (MOCP) can be approximated by the following parameter optimization problem:

$$\text{(MOCP(p))} \quad \min\ \tilde{J}(\sigma^p) = -x_3(T|\sigma^p)$$
$$\text{s.t. } \sigma^p \in \mathcal{F}^p.$$

We relate the optimal control of (MOCP) and the optimal parameter vector of (MOCP(p)) in the following theorems. The proofs of these theorems are similar to that given for Theorems 6.5.1 and 6.5.2 in [240].

**Theorem 5.4.** *Let $\sigma^{p,*}$ be an optimal parameter vector of* (MOCP(p)) *and $u^{p,*}$ be the corresponding optimal control constructed by* (5.22). *Suppose that $u^*$ is an optimal control of* (MOCP). *Then*

$$\lim_{\min_{i\in\Lambda}\{n_{p_i}\}\to\infty} J(u^{p,*}) = J(u^*).$$

**Theorem 5.5.** *Let $\sigma^{p,*}$ and $u^{p,*}$ be defined as in Theorem* 5.4. *Suppose that*

$$\lim_{\min_{i\in\Lambda}\{n_{p_i}\}\to\infty} u^{p,*}(t) = \bar{u}(t)\ a.e.\ in\ [0, T].$$

*Then, $\bar{u}$ is an optimal control of* (MOCP).

To solve (MOCP) by the control parameterization method, we need to solve a sequence of problems $\{\text{MOCP(p)}\}_{p=1}^{\infty}$. However, it is difficult to cope with the continuous state inequality constraint (5.25). To surmount these difficulties, let

$$g_\ell(x(t|\sigma^p)) = x_\ell(t|\sigma^p) - x_\ell^*,$$

$$g_{5+\ell}(\boldsymbol{x}(t|\boldsymbol{\sigma}^p)) = x_{*\ell} - x_\ell(t|\boldsymbol{\sigma}^p),\ \ell = 1, 2, \ldots, 5.$$

The condition $\boldsymbol{x}(t|\boldsymbol{\sigma}^p) \in W$ is equivalently transcribed into

$$G(\boldsymbol{\sigma}^p) = 0, \tag{5.26}$$

where $G(\boldsymbol{\sigma}^p) := \sum_{l=1}^{10} \int_0^T \max\{0, g_l(\boldsymbol{x}(t|\boldsymbol{\sigma}^p))\}^2 \mathrm{d}t$.

Then, the gradient of the constraint $G(\cdot)$ can be computed by the following theorem.

**Theorem 5.6.** *For the constraint $G(\boldsymbol{\sigma}^p)$ given in (5.26), it holds that its gradient with respect to parameterized control $\boldsymbol{\sigma}^p$ is*

$$\frac{\partial G(\boldsymbol{\sigma}^p)}{\partial \boldsymbol{\sigma}^p} = \int_0^T \frac{\partial H(t, \boldsymbol{x}(t), \boldsymbol{\sigma}^p, \boldsymbol{\lambda}(t))}{\partial \boldsymbol{\sigma}^p} \mathrm{d}t, \tag{5.27}$$

*where*

$$H(t, \boldsymbol{x}(t), \boldsymbol{\sigma}^p, \boldsymbol{\lambda}(t)) = \sum_{l=1}^{10} \max\{0, g_l(\boldsymbol{x}(t|\boldsymbol{\sigma}^p))\}^2 + \boldsymbol{\lambda}^\top(t)\tilde{\boldsymbol{f}}(t, \boldsymbol{x}(t), \boldsymbol{\sigma}^p),$$

*and*

$$\boldsymbol{\lambda}(t) = (\lambda_1(t), \lambda_2(t), \lambda_3(t), \lambda_4(t), \lambda_5(t))^\top$$

*is the solution of the costate system*

$$\dot{\boldsymbol{\lambda}}(t) = -\left(\frac{\partial H(t, \boldsymbol{x}(t), \boldsymbol{\sigma}^p, \boldsymbol{\lambda}(t))}{\partial \boldsymbol{x}}\right)^\top,$$

*with the boundary conditions*

$$\boldsymbol{\lambda}(T) = (0, 0, 0, 0, 0)^\top,$$
$$\boldsymbol{\lambda}(0) = (0, 0, 0, 0, 0)^\top,$$
$$\boldsymbol{\lambda}(t_i+) = \boldsymbol{\lambda}(t_i-), \quad i = 1, 2, \ldots, 2N.$$

*Proof.* The proof can be completed using the method of Chapter 3 in [38].

Based on the above theorems, each of these {MOCP(p)} can be viewed as a constrained mathematical programming problem, which can be solved by various optimization methods such as gradient-based techniques [242]. However, all those techniques are only designed to find local optimal solutions.

PSO is an evolutionary computational method which is based on swarm intelligence. Presently, PSO has attracted wide attention in evolutionary computing, optimization, and many other fields [98, 132, 280]. In a typical PSO, each particle "flies" over the search space to look for promising regions according to the experiences of both its own and those of the groups. Thus, the social sharing of information takes place and individuals profit from the discoveries and previous experiences of all other particles in a wide landscape during their search process around the better solutions. Traditionally, the original PSO method deals with unconstrained optimization problems. However, what we need to solve is an optimization problem with both control bounds and state constraints, to which the original PSO cannot be applied directly. Moreover, the original PSO is easy to converge to the local optima. So we make some improvements to the original PSO proposed in [117]. In order to handle the bounds of control in (MOCP(p)), a reflection strategy is introduced. The constraints of state are very difficult to handle. Here, the gradients of constraints are utilized. On the basis of Theorem 5.6, we propose a strategy of dealing with state constraints. A new updating strategy about velocity and position and a craziness operator to overcome local convergence are also introduced. Considering $N^p$ particles in the evolution process, the position and velocity of the $i$th particle can be represented by $\boldsymbol{\sigma}_i^p := \left(\sigma_{i,1}^p, \sigma_{i,2}^p, \ldots, \sigma_{i,\kappa}^p\right)^\top$ and $\boldsymbol{v}_i^p := \left(v_{i,1}^p, v_{i,2}^p, \ldots, v_{i,\kappa}^p\right)^\top$, respectively. Furthermore, denote the lower bound and upper bound of the position by $\boldsymbol{\sigma}_{low}$ and $\boldsymbol{\sigma}_{upp}$, which can be obtained by (5.23), respectively. At the $(k+1)$th iteration, the improved evolutionary strategies of particle $i$ are as follows.

1. (Velocity and position updating) In the former stage of iterations, velocity and position are updated by the following modifications to balance exploration (global investigation of the parameter space) and exploitation (the refinement of searches around a local optimum):

$$v_{i,j}^p(k+1) := r_{ij}^3 v_{i,j}^p(k) + c_1^p r_{ij}^1 \left(pb_{i,j}^p - \sigma_{i,j}^p(k)\right) + c_2^p r_{ij}^2 \left(gb_j^p - \sigma_{i,j}^p(k)\right),$$

$$\sigma_{i,j}^p(k+1) := r_{ij}^4 \sigma_{i,j}^p(k) + \left(1 - r_{ij}^4\right) v_{i,j}^p(k+1),$$

   where $\boldsymbol{pb}_i^p := \left(pb_{i,1}^p, pb_{i,2}^p, \ldots, pb_{i,\kappa}^p\right)^\top$ is the best position that particle $i$ has ever found, $\boldsymbol{gb}^p := \left(gb_1^p, gb_2^p, \ldots, gb_\kappa^p\right)^\top$ is the best position that the group has ever found, $c_1^p$ and $c_2^p$ are two positive constants, and $r_{ij}^1, r_{ij}^2, r_{ij}^3$, and $r_{ij}^4$ are random parameters chosen uniformly from [0, 1].
2. (Craziness operator) When the number of iterations is larger than $M_1^p$, the velocity of the $i$th particle is adjusted, to keep the diversity of particles, as

$$v_{i,j}^p(k+1) := \begin{cases} v_{i,j}^p(k+1), & \text{if } r_5 \leqslant P_{cr}, \\ 2\mathrm{Rand}_{i,j}(r_5) - 1, & \text{otherwise,} \end{cases}$$

where $r_5$ is a random parameter which is taken uniformly from [0, 1], $\text{Rand}_{i,j}(\cdot)$ is a function which is used to randomly generate the $j$th component of the velocity of the $i$th particle, and $P_{cr}$ is a predefined probability.

3. (Dealing with position outside control bounds) Assume that the $j$th component of position of the $i$th particle at the $(k+1)$th step violates boundary constraints, then it is reflected back from the bound by the amount of violation:

$$\sigma_{i,j}^{p}(k+1) := \begin{cases} 2\sigma_{\text{low},j} - \sigma_{i,j}^{p}(k+1), & \text{if } \sigma_{i,j}^{p}(k+1) < \sigma_{\text{low},j}, \\ 2\sigma_{\text{upp},j} - \sigma_{i,j}^{p}(k+1), & \text{if } \sigma_{i,j}^{p}(k+1) > \sigma_{\text{upp},j}. \end{cases}$$

4. (Strategy of dealing with state constraints) For the position of the $i$th particle at the $(k+1)$th step, test the value of $G\left(\boldsymbol{\sigma}_i^p(k+1)\right)$. If $G\left(\boldsymbol{\sigma}_i^p(k+1)\right) = 0$, then the position is feasible. Otherwise, that is, $G\left(\boldsymbol{\sigma}_i^p(k+1)\right) > 0$, move the position toward the feasible region in the direction of $-\dfrac{\partial G\left(\boldsymbol{\sigma}_i^p(k+1)\right)}{\partial \boldsymbol{\sigma}_i^p(k+1)}$ with Armijo line search.
5. (Stopping criteria) The algorithm stops when any of the following conditions holds:

   - The maximal iteration $M^p$ is reached;
   - The maximal deviation between the group's best fitness values in the last $M_2^p$ iterations is less than $\varepsilon^p$, where $\varepsilon^p$ is a predefined constant.

On the basis of the above improved PSO algorithm, we can obtain an approximately optimal control for (MOCP) as shown in the following algorithm.

**Algorithm 5.1.**

**Step 1.** Solve (MOCP(p)) using the improved PSO algorithm to compute $\boldsymbol{\sigma}^{p,*}$.
**Step 2.** If $\min_{i \in \Lambda} n_{p_i} \geqslant P$, where $P$ is a predefined positive constant, then go to Step 3. Otherwise go to Step 1 with $n_{p_i}$ increased to $n_{p_i+1}$ for each $i$.
**Step 3.** Construct $u^{p,*}$ from $\boldsymbol{\sigma}^{p,*}$ such that

$$u^{p,*}\left(t|\boldsymbol{\sigma}^{p,*}\right) = \sum_{i=1}^{2N+1} \sum_{k=1}^{n_{p_i}} \sigma^{p_i,*,k} \chi_{\left(\tau_{k-1}^{p_i,*}, \tau_k^{p_i,*}\right]}(t),\ t \in [0, T], \tag{5.28}$$

and stop.

The piecewise constant control $u^{p,*}$ obtained is an approximately optimal solution of (MOCP).

## 5.6 Numerical Results

In the numerical simulation, the initial state, initial volume of fermentor, velocity ratio of adding alkali to glycerol, concentration of initial feeding of glycerol, and fermentation time are $\boldsymbol{x}_0 = (0.1115\,\mathrm{g\,L^{-1}}, 495\,\mathrm{mmol\,L^{-1}}, 0, 0, 0)^{\top}$, $V_0 = 5\,\mathrm{L}$, $r = 0.75$, $c_{s0} = 10{,}762\,\mathrm{mmol\,L^{-1}}$, and $T = 24.16\,\mathrm{h}$, respectively. Fed-batch process begins at $t_1 = 5.33\,\mathrm{h}$. The feeding moment $t_{2j+1}$, the feeding stopping moment $t_{2j+2}$, and $j \in \bar{\Lambda}_1 = \{0, 1, \ldots, 676\}$ are determined by the experiment. In order to save computational time, the fermentation process is partitioned into the first batch phase (Bat. Ph.) and phases I–IX (Phs. I–IX) according to the actual experiment. In each one of Phs. I–IX, the same feeding strategy is adopted. Moreover, the durations of the feeding processes in Phs. I–IX are 5, 7, 8, 7, 6, 4, 3, 2, and 1 s in each 100 s, leaving 95, 93, 92, 93, 94, 96, 97, 98, and 99 s for batch cultures, respectively. The bounds of feeding rates in Phs. I–IX are listed in Table 5.2.

In the improved PSO algorithm, the number of initial particles swarm $N^p$, the maximal iteration $M^p$, and the parameters $c_1^p$, $c_2^p$, $P_{cr}$, $M_1^p$, $M_2^p$, and $\varepsilon_p$ are, respectively, 200, 100, 2, 2, 0.5, 50, 20, and $10^{-3}$. These parameters are derived empirically after numerous experiments. In Algorithm 5.1, $P$ takes value 1, and the ODEs are numerically calculated by improved Euler method with the relative error tolerance $10^{-4}$. Applying Algorithm 5.1 to the optimal control model, we obtain the optimal feeding strategy of glycerol. It takes about 53.6 s to iterate one step on an AMD Athlon 64 X2 Dual Core Processor TK-57 1.90 GHz machine. The feeding rates in Phs. I–IX are shown in Fig. 5.1. Furthermore, we obtain that the concentration of 1,3-PD at the terminal time is $925.127\,\mathrm{mmol\,L^{-1}}$, which is increased by 16.04 % in comparison with $797.23\,\mathrm{mmol\,L^{-1}}$ in the experiment. The concentration change of 1,3-PD obtained by the optimal feeding strategy is shown in Fig. 5.2. The curve also confirms that the simulation result is better than the one in the experiment.

**Table 5.2** The bounds of feeding rates in Phs. I–IX

| | Phs. I–II Phs. IV–V | Ph. III | Ph. VI | Ph. VII | Phs. VIII–IX |
|---|---|---|---|---|---|
| Upper bounds | 0.2524 | 0.2390 | 0.2657 | 0.2924 | 0.3058 |
| Lower bounds | 0.1682 | 0.1594 | 0.1771 | 0.1949 | 0.2038 |

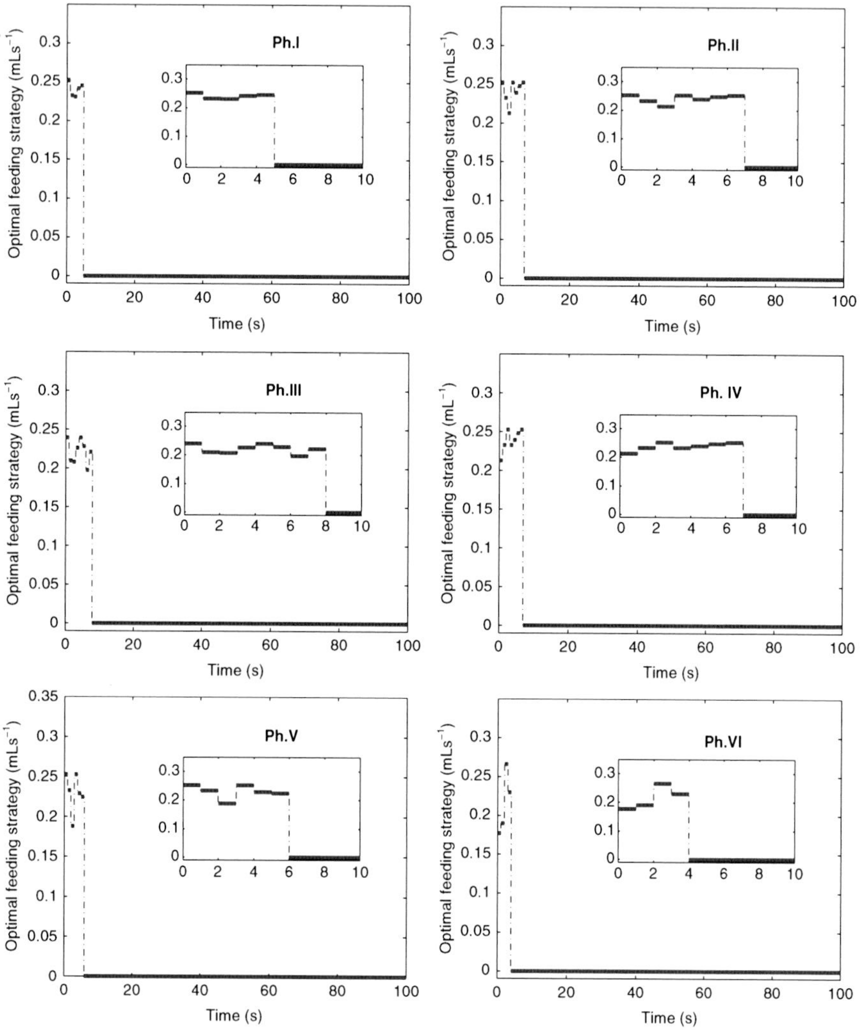


**Fig. 5.1** The optimal feeding strategy of glycerol in fed-batch fermentation process

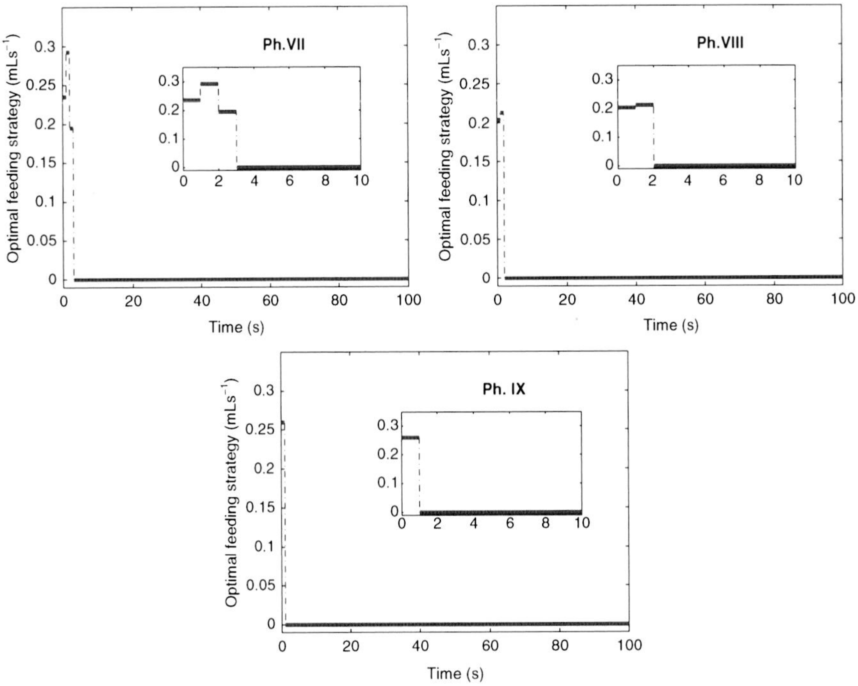


**Fig. 5.1** (continued)

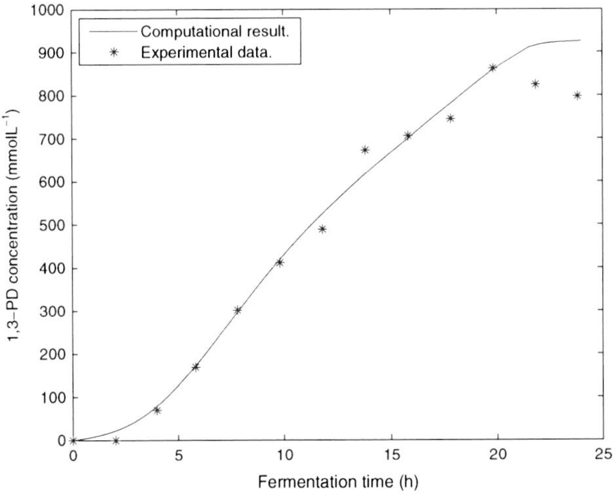


**Fig. 5.2** The concentration changes of 1,3-PD with respect to fermentation time

## 5.7 Conclusion

In this chapter, we presented a controlled nonlinear multistage dynamical system to formulate the fed-batch process. The multistage optimal control problem was investigated. A global optimization algorithm was constructed to solve the optimal control problem. Numerical results show that the target product concentration could be increased considerably.

# Chapter 6
# Optimal Control of Switched Autonomous Systems

## 6.1 Introduction

In this chapter, we consider optimal control of a switched autonomous system arising in constantly fed-batch fermentation. Constantly fed-batch fermentation, a simple mode with a constant feeding rate, has been widely applied for the production of many bioproducts [4, 93, 288]. For glycerol bioconversion to 1,3-PD in constantly fed-batch fermentation, we model this process as a switched autonomous system. Then, the optimal control problem, in which the switching instants are taken as the control function, involving the switched autonomous system and subject to state constraints and control constraint is investigated. Incidentally, optimal control of switched autonomous systems has been extensively discussed [23, 32, 72, 87, 187, 273]. Nevertheless, to our knowledge, optimal control of switched autonomous systems with continuous state constraints has rarely been considered.

In this chapter, by applying a time-scaling transformation, we obtain an equivalent problem with parameters and fixed switching instants. The existence of the optimal control is also proved. A computational approach is developed to seek the optimal solution in three aspects. Firstly, we transcribe the continuous state inequality constraints into an equality constraint by a constraint transcription. Secondly, the local smoothing technique in [243] is utilized to approximate the non-smooth equality constraint, and the convergence of this approximate approach is also presented. Thirdly, a penalty function that incorporates the transformed constraint to form a new cost function is introduced [189]. Numerical results illustrate the effectiveness of the proposed algorithm.

The main references of this chapter are [91] and [146].

 
C. Liu, Z. Gong, *Optimal Control of Switched Systems Arising in Fermentation Processes*, Springer Optimization and Its Applications 97,
DOI 10.1007/978-3-662-43793-3_6

## 6.2 Switched Autonomous Systems

In the fed-batch process, the composition of culture medium, cultivation conditions, and analytical methods of fermentative products were similar to those previously reported [48]. According to the fermentation process, we assume that

**Assumption 6.1.** *During the process of fed-batch culture, only glycerol and alkali are fed into the reactor with certain constant velocity. Moreover, the feeding velocity ratio of alkali to glycerol is a constant.*

Let $\boldsymbol{x}(t) := (x_1(t), x_2(t), x_3(t), x_4(t), x_5(t), x_6(t))^\top \in \mathbb{R}^6_+, t \in [0, T]$, be the state vector. The components of $\boldsymbol{x}(t)$ represent the extracellular concentrations of biomass, glycerol, 1,3-PD, acetate, and ethanol concentrations and the volume of culture fluid at $t$ in fermentor, respectively. $T$ is the terminal time of the fermentation and $\boldsymbol{x}_0$ is a given initial state. Let $\tau_i, i \in \Lambda := \{1, 2, \ldots, 2N+1\}$, be the switching instants such that $0 = \tau_0, \tau_{i-1} < \tau_i, i \in \Lambda$, and $\tau_{2N+1} = T$. Here, $N$ is a constant in this chapter. In particular, $\tau_{2j+1}$ is the moment of adding glycerol, and $\tau_{2j+2}$ denotes the moment of ending the flow of glycerol, $j \in \bar{\Lambda}_1 := \{0, 1, 2, \ldots, N-1\}$. Under Assumptions 5.1 and 6.1, mass balances of biomass, substrate, and products in fed-batch culture can be formulated as the following switched autonomous system

$$\begin{cases} \dot{\boldsymbol{x}}(t) = \boldsymbol{f}^i(\boldsymbol{x}(t)), \\ \boldsymbol{x}(\tau_{i-1}+) = \boldsymbol{x}(\tau_{i-1}),\ t \in (\tau_{i-1}, \tau_i],\ i = 1, 2, \ldots, 2N+1, \\ \boldsymbol{x}(0) = \boldsymbol{x}_0, \end{cases} \tag{6.1}$$

where, for $t \in (\tau_{2j}, \tau_{2j+1}], j \in \bar{\Lambda}_2 := \{0, 1, \ldots, N\}$,

$$\boldsymbol{f}^{2j+1}(\boldsymbol{x}(t)) = \begin{pmatrix} q_1(\boldsymbol{x}(t))x_1(t) \\ -q_2(\boldsymbol{x}(t))x_1(t) \\ q_3(\boldsymbol{x}(t))x_1(t) \\ q_4(\boldsymbol{x}(t))x_1(t) \\ q_5(\boldsymbol{x}(t))x_1(t) \\ 0 \end{pmatrix}; \tag{6.2}$$

for $t \in (\tau_{2j+1}, \tau_{2j+2}], j \in \bar{\Lambda}_1$,

$$\boldsymbol{f}^{2j+2}(\boldsymbol{x}(t)) = \begin{pmatrix} (q_1(\boldsymbol{x}(t)) - D(\boldsymbol{x}(t)))x_1(t) \\ D(\boldsymbol{x}(t))\left(\dfrac{c_{s0}}{1+r} - x_2(t)\right) - q_2(\boldsymbol{x}(t))x_1(t) \\ q_3(\boldsymbol{x}(t))x_1(t) - D(\boldsymbol{x}(t))x_3(t) \\ q_4(\boldsymbol{x}(t))x_1(t) - D(\boldsymbol{x}(t))x_4(t) \\ q_5(\boldsymbol{x}(t))x_1(t) - D(\boldsymbol{x}(t))x_5(t) \\ (1+r)v \end{pmatrix}. \tag{6.3}$$

In (6.3), $r$ is the velocity ratio of adding alkali to glycerol, $c_{s0}$ denotes the initial concentration of glycerol in feed, and $D(\boldsymbol{x}(t))$ is the dilution rate defined by

$$D(\boldsymbol{x}(t)) = \frac{(1+r)v}{x_6(t)}. \tag{6.4}$$

In (6.3)–(6.4), $v > 0$ is the velocity of feeding glycerol and is a constant. As in Chap. 5, the specific growth rate of cells $q_1(\boldsymbol{x}(t))$, the specific consumption rate of substrate $q_2(\boldsymbol{x}(t))$, and the specific formation rates of products $q_\ell(\boldsymbol{x}(t))$, $\ell = 3, 4, 5$, are expressed by

$$q_1(\boldsymbol{x}(t)) = \frac{\Delta_1 x_2(t)}{x_2(t) + k_1} \prod_{\ell=2}^{5} \left(1 - \frac{x_\ell(t)}{x_\ell^*}\right)^{n_\ell}, \tag{6.5}$$

$$q_2(\boldsymbol{x}(t)) = m_2 + q_1(\boldsymbol{x}(t))Y_2 + \frac{\Delta_2 x_2(t)}{x_2(t) + k_2}, \tag{6.6}$$

$$q_\ell(\boldsymbol{x}(t)) = -m_\ell + q_1(\boldsymbol{x}(t))Y_\ell + \frac{\Delta_\ell x_2(t)}{x_2(t) + k_\ell}, \quad \ell = 3, 4, \tag{6.7}$$

$$q_5(\boldsymbol{x}(t)) = q_2(\boldsymbol{x}(t)) \left( \frac{c_1}{c_2 + q_1(\boldsymbol{x}(t))x_2(t)} + \frac{c_3}{c_4 + q_1(\boldsymbol{x}(t))x_2(t)} \right). \tag{6.8}$$

Under anaerobic conditions at 37 °C and pH 7.0, the critical concentrations for cell growth and the parameters in (6.5)–(6.8) are as given in Table 5.1.

Since biological considerations limit the rate of switching, there are maximal and minimal time durations that are spent on each one of the batch process and the feeding process. On this basis, define the set of admissible switching instants as

$$\begin{aligned} \Gamma := \{(\tau_1, \tau_2, \ldots, \tau_{2N})^\top \in \mathbb{R}^{2N} \mid \rho_i \leqslant \tau_i - \tau_{i-1} \leqslant \delta_i, i = 1, 2, \ldots, 2N+1, \\ \tau_1 + N \cdot (\tau_{2\iota+1} - \tau_{2\iota-1}) = T, \iota = 1, 2, \ldots, N\}, \end{aligned} \tag{6.9}$$

where $\tau_0 = 0$, $\tau_{2N+1} = T$, $\rho_i$ and $\delta_i$ are the minimal and maximal time durations of the $i$th process, respectively. Accordingly, any $\boldsymbol{\tau} \in \Gamma$ is regarded as an admissible vector of switching instants. Moreover, the concentrations of biomass, glycerol, and products and the volume of culture fluid are restricted in a certain range according to the fermentation process, so we consider the properties of the system (6.1) within $\tilde{W} := \prod_{\ell=1}^{6} [x_{*\ell}, x_\ell^*]$, where $x_{*\ell}$, $x_\ell^*$, $\ell = 1, 2, \ldots, 5$, are as given in Table 5.1, $x_{*6} = 4$ and $x_6^* = 7$.

For the system (6.1), some important properties are given in the following theorems.

**Theorem 6.1.** *The functions* $\boldsymbol{f}^i(\cdot), i = 1, 2, \ldots, 2N+1$, *defined in* (6.2) *and* (6.3) *satisfy that*

(a) $\boldsymbol{f}^i(\cdot) : \mathbb{R}^6_+ \to \mathbb{R}^6$, *together with their partial derivatives with respect to* $\boldsymbol{x}$, *are continuous on* $\mathbb{R}^6_+$; *and*
(b) *There exists a constant* $K > 0$ *such that the linear growth condition*

$$\max\{\|\boldsymbol{f}^i(\boldsymbol{x})\| \mid i = 1, 2, \ldots, 2N+1\} \leqslant K(1 + \|\boldsymbol{x}\|), \forall \boldsymbol{x} \in \mathbb{R}^6_+ \quad (6.10)$$

*holds, where* $\|\cdot\|$ *is the Euclidean norm.*

*Proof.* (a) This conclusion can be obtained by the expressions of $\boldsymbol{f}^i$ in (6.2)–(6.3).
(b) We can prove this inequality in a similar manner to the proof of Proposition 5.1 in Chap. 5. □

**Theorem 6.2.** *For each* $\boldsymbol{\tau} \in \Gamma$, *the switched autonomous system* (6.1) *has a unique continuous solution denoted by* $\boldsymbol{x}(\cdot|\boldsymbol{\tau})$. *Furthermore,* $\boldsymbol{x}(\cdot|\boldsymbol{\tau})$ *satisfies the following integral equation*

$$\boldsymbol{x}(t|\boldsymbol{\tau}) = \boldsymbol{x}(\tau_{i-1}|\boldsymbol{\tau}) + \int_{\tau_{i-1}}^{t} \boldsymbol{f}^i(\boldsymbol{x}(s|\boldsymbol{\tau}))\mathrm{d}s, \forall t \in (\tau_{i-1}, \tau_i],\ i = 1, 2, \ldots, 2N+1, \quad (6.11)$$

*and is continuous in* $\boldsymbol{\tau}$.

*Proof.* This conclusion can be obtained from Theorem 6.1 and the theory of ordinary differential equations [5]. □

## 6.3 Optimal Control Models

For mathematical convenience, define the set of the solutions to the system (6.1), $\mathcal{S}_0$, as follows.

$$\mathcal{S}_0 = \{\boldsymbol{x}(\cdot|\boldsymbol{\tau}) | \ \boldsymbol{x}(t|\boldsymbol{\tau}) \text{ is the continuous solution to the system (6.1) with } \boldsymbol{\tau} \in \Gamma \text{ for all } t \in [0, T]\}.$$

Since the concentrations of biomass, glycerol, and products are restricted in $\tilde{W}$, we denote the set of the admissible solutions by

$$\mathcal{S} = \{\boldsymbol{x}(\cdot|\boldsymbol{\tau}) \in \mathcal{S}_0 | \ \boldsymbol{x}(t|\boldsymbol{\tau}) \in \tilde{W} \text{ for all } t \in [0, T]\}. \quad (6.12)$$

Furthermore, the set of the feasible vectors of switching instant can be defined as

$$F = \{\boldsymbol{\tau} \in \Gamma | \ \boldsymbol{x}(\cdot|\boldsymbol{\tau}) \in \mathcal{S}\}. \quad (6.13)$$

In the fermentation process of glycerol to 1,3-PD, it is desired that the 1,3-PD concentration should be maximized at the end of the process by optimizing the switching instants between the batch process and the feeding process. Thus, the optimal control problem can now be described as follows:

$$\text{(AOCP)} \qquad \min\ J(\boldsymbol{\tau}) = -x_3(T|\boldsymbol{\tau})$$
$$\text{s.t. } \boldsymbol{\tau} \in F,$$

where $x_3(\cdot|\boldsymbol{\tau})$ is the third component of the solution to the system (6.1).

It is difficult to solve the switched optimal control model (AOCP) using existing numerical techniques. The main difficulty is the implicit dependence of the system state on the variable switching instants. We now employ a time-scaling transformation to map these switching instants into a fixed set of time points in a new time horizon.

Define

$$\begin{aligned}\Theta := \{\boldsymbol{\theta} \in \mathbb{R}^{2N+1} \mid \rho_i \leqslant \theta_i \leqslant \delta_i, i = 1, 2, \ldots, 2N+1,\\ \theta_1 + N \cdot (\theta_{2\iota+1} + \theta_{2\iota}) = T, \iota = 1, 2, \ldots, N\}.\end{aligned} \tag{6.14}$$

Let $s \in [0, 2N+1]$ be a new time variable with switching instants occurring at the fixed points $s = \varsigma, \varsigma = 1, 2, \ldots, 2N$. For each $\boldsymbol{\theta} \in \Theta$, define $\mu(\cdot|\boldsymbol{\theta}) : [0, 2N+1] \to \mathbb{R}$ by

$$\mu(s|\boldsymbol{\theta}) = \begin{cases} \sum_{i=1}^{\lfloor s \rfloor} \theta_i + \theta_{\lfloor s \rfloor + 1}(s - \lfloor s \rfloor), & \text{if } s \in [0, 2N+1), \\ T, & \text{if } s = 2N+1, \end{cases} \tag{6.15}$$

where $\lfloor \cdot \rfloor$ denotes the floor function. It can be easily verified that $\mu(\cdot|\boldsymbol{\theta})$ is continuous and strictly increasing on $[0, 2N+1]$. Consequently, $\mu(\cdot|\boldsymbol{\theta}) : [0, 2N+1] \to [0, T]$ is a bijection. Under this mapping, the new uniform switching instants are mapped to the following value in the original time scale:

$$\mu(i|\boldsymbol{\theta}) = \sum_{k=1}^{i} \theta_k, i = 1, \ldots, 2N+1. \tag{6.16}$$

Let $\tilde{\boldsymbol{x}}(s) := \boldsymbol{x}(\mu(s|\boldsymbol{\theta}))$ and $\boldsymbol{h}^i(\tilde{\boldsymbol{x}}(s), \boldsymbol{\theta}) := \theta_i \boldsymbol{f}^i(\boldsymbol{x}(\mu(s|\boldsymbol{\theta})))$. Then, the system (6.1) becomes

$$\begin{cases} \dot{\tilde{\boldsymbol{x}}}(s) = \boldsymbol{h}^i(\tilde{\boldsymbol{x}}(s), \boldsymbol{\theta}), \\ \tilde{\boldsymbol{x}}(i-1+) = \tilde{\boldsymbol{x}}(i-1),\ s \in (i-1, i],\ i = 1, 2, \ldots, 2N+1, \\ \tilde{\boldsymbol{x}}(0) = \boldsymbol{x}_0, \end{cases}$$

where $\tilde{\boldsymbol{x}}(i-1+)$ denotes the right limit of $\tilde{\boldsymbol{x}}(s)$ at $i-1$.

Let $\tilde{\boldsymbol{x}}(\cdot|\boldsymbol{\theta})$ be the continuous solution of the system (6.17) corresponding to $\boldsymbol{\theta} \in \Theta$. Then, the set of feasible switching instants (6.13) turns into

$$\tilde{F} = \{\boldsymbol{\theta} \in \Theta | \; \tilde{\boldsymbol{x}}(s|\boldsymbol{\theta}) \in \tilde{W} \text{ for all } s \in [0, 2N + 1]\}. \tag{6.17}$$

Thus, (AOCP) is equivalently transcribed into the following parametrization optimal control problem with fixed switching instants:

$$\text{(PAOCP)} \qquad \min \; \tilde{J}(\boldsymbol{\theta}) = -\tilde{x}_3(2N + 1|\boldsymbol{\theta})$$
$$\text{s.t. } \; \boldsymbol{\theta} \in \tilde{F}.$$

Now, the existence theorem of the optimal solution to (AOCP) is stated as follows.

**Theorem 6.3.** (AOCP) *has at least one optimal solution.*

*Proof.* Note that (AOCP) is equivalent to the (PAOCP). Consequently, it suffices to consider the existence of optimal solution to (PAOCP). In view of the compactness of the set $\Theta$, we obtain that $\tilde{F}$ is a bounded set. Then, for any sequence $\left\{\boldsymbol{\theta}^k\right\}_{k=1}^{\infty} \subseteq \Theta$, there exists at least one subsequence $\left\{\hat{\boldsymbol{\theta}}^{k_j}\right\} \subseteq \left\{\boldsymbol{\theta}^k\right\}$ such that $\hat{\boldsymbol{\theta}}^{k_j} \to \hat{\boldsymbol{\theta}}$ as $j \to \infty$. It follows from Theorem 6.2 and the property of $\mu(\cdot|\boldsymbol{\theta})$ that the solution $\tilde{\boldsymbol{x}}(\cdot|\boldsymbol{\theta})$ of (6.17) is continuous in $\boldsymbol{\theta}$. Now, suppose $\tilde{\boldsymbol{x}}\left(s|\hat{\boldsymbol{\theta}}^{k_j}\right) \in \tilde{W}$ for all $s \in [0, 2N + 1]$, then $\tilde{\boldsymbol{x}}\left(\cdot|\hat{\boldsymbol{\theta}}\right)$ is a solution of the system (6.17) and $\tilde{\boldsymbol{x}}\left(s|\hat{\boldsymbol{\theta}}\right) \in \tilde{W}$ for all $s \in [0, 2N + 1]$ due to the compactness of $\tilde{W}$. That is, $\hat{\boldsymbol{\theta}} \in \tilde{F}$, which implies the closeness of $\tilde{F}$. Furthermore, since the cost function $\tilde{J}(\boldsymbol{\theta})$ is continuous in $\boldsymbol{\theta}$, we obtain that (PAOCP) has at least one optimal solution. □

## 6.4 Computational Approaches

(PAOCP) is essentially an optimization problem subject to continuous state constraints. It can be viewed as a semi-infinite programming problem. An efficient algorithm for solving optimization problems of this type is discussed in [243]. We will now briefly discuss the application of this algorithm to (PAOCP).

Let

$$g_\ell(\tilde{\boldsymbol{x}}(s|\boldsymbol{\theta})) = \tilde{x}_\ell(s|\boldsymbol{\theta}) - x_\ell^*,$$
$$g_{6+\ell}(\tilde{\boldsymbol{x}}(s|\boldsymbol{\theta})) = x_{*\ell} - \tilde{x}_\ell(s|\boldsymbol{\theta}), \; \ell = 1, 2, \ldots, 6.$$

Then, the condition $\tilde{\boldsymbol{x}}(s|\boldsymbol{\theta}) \in \tilde{W}$ is equivalently transcribed into

$$G(\boldsymbol{\theta}) = 0, \tag{6.18}$$

where $G(\boldsymbol{\theta}) := \sum_{l=1}^{12} \int_0^{2N+1} \max\{0, g_l(\tilde{\boldsymbol{x}}(s|\boldsymbol{\theta}))\} \mathrm{d}s$. As stated in [243], this constraint transcription satisfies the constraint qualification (3.7). However, since $G(\boldsymbol{\theta})$ is non-smooth in $\boldsymbol{\theta}$, standard optimization routines would have difficulties in dealing with this type of equality constraints. The following smoothing technique is to replace $\max\{0, g_l(\tilde{\boldsymbol{x}}(s|\boldsymbol{\theta}))\}$ with $\hat{g}_{l,\epsilon}(\tilde{\boldsymbol{x}}(s|\boldsymbol{\theta}))$, where

$$\hat{g}_{l,\epsilon}(\tilde{\boldsymbol{x}}(s|\boldsymbol{\theta})) = \begin{cases} 0, & \text{if } g_l(\tilde{\boldsymbol{x}}(s|\boldsymbol{\theta})) < -\epsilon, \\ \dfrac{(g_l(\tilde{\boldsymbol{x}}(s|\boldsymbol{\theta})) + \epsilon)^2}{4\epsilon}, & \text{if } -\epsilon \leqslant g_l(\tilde{\boldsymbol{x}}(s|\boldsymbol{\theta})) \leqslant \epsilon, \\ g_l(\tilde{\boldsymbol{x}}(s|\boldsymbol{\theta})), & \text{if } g_l(\tilde{\boldsymbol{x}}(s|\boldsymbol{\theta})) > \epsilon. \end{cases} \tag{6.19}$$

Note that

$$G_\epsilon(\boldsymbol{\theta}) = \sum_{l=1}^{12} \int_0^{2N+1} \hat{g}_{l,\epsilon}(\tilde{\boldsymbol{x}}(s|\boldsymbol{\theta}))\mathrm{d}s \tag{6.20}$$

is a smooth function in $\boldsymbol{\theta}$. Let

$$\begin{aligned} \tilde{F}_\epsilon &= \{\boldsymbol{\theta} \in \Theta |\ G_\epsilon(\boldsymbol{\theta}) = 0\} \\ &= \{\boldsymbol{\theta} \in \Theta |\ g_l(\tilde{\boldsymbol{x}}(s|\boldsymbol{\theta})) \leqslant -\epsilon, l = 1, 2, \ldots, 12, s \in [0, 2N+1]\}. \end{aligned} \tag{6.21}$$

Clearly, $\tilde{F}_\epsilon \subset \tilde{F}$ for each $\epsilon > 0$.

We now define an approximate problem, denoted by $\left(\text{PAOCP}_{\epsilon,\gamma}\right)$, where the smoothed state constraints are treated as a penalty function.

$$\left(\text{PAOCP}_{\epsilon,\gamma}\right) \qquad \begin{aligned} &\min\ \tilde{J}_{\epsilon,\gamma}(\boldsymbol{\theta}) = -\tilde{x}_3(2N+1|\boldsymbol{\theta}) + \gamma G_\epsilon(\boldsymbol{\theta}) \\ &\text{s.t. } \boldsymbol{\theta} \in \Theta. \end{aligned} \tag{6.22}$$

We present the relationship between the solution of (PAOCP) and that of $\left(\text{PAOCP}_{\epsilon,\gamma}\right)$ in the following theorems. The proofs are similar to those given for Theorems 2.1 and 2.2 in [243].

**Theorem 6.4.** *There exists a $\gamma(\epsilon) > 0$ such that for all $\gamma > \gamma(\epsilon)$ any solution to $\left(\text{PAOCP}_{\epsilon,\gamma}\right)$ is also a feasible point of* (PAOCP).

**Theorem 6.5.** *Let $\boldsymbol{\theta}^*$ be an optimal solution to* (PAOCP) *and $\boldsymbol{\theta}^*_{\epsilon,\gamma}$ an optimal solution to $\left(\text{PAOCP}_{\epsilon,\gamma}\right)$, where $\gamma$ is chosen appropriately to ensure that $\boldsymbol{\theta}^*_{\epsilon,\gamma} \in \tilde{F}$. Then*

$$\lim_{\epsilon \to 0} \tilde{J}\left(\boldsymbol{\theta}^*_{\epsilon,\gamma}\right) = \tilde{J}\left(\boldsymbol{\theta}^*\right). \tag{6.23}$$

On this basis, (AOCP) can be solved by solving a sequence of approximate problems $\{(\text{PAOCP}_{\epsilon,\gamma})\}$. Each of these $\{(\text{PAOCP}_{\epsilon,\gamma})\}$ is a smooth nonlinear mathematical programming problem, which can be solved by various optimization methods [30, 189, 240]. In particular, the sequential quadratic programming (SQP) is one of effective methods because of its superlinear convergence [30]. For this, we need the gradient formulae of the cost function (6.22) with respect to parameter vector $\boldsymbol{\theta}$. These gradient formulae are presented in the next theorem.

**Theorem 6.6.** *Consider the* $(\text{PAOCP}_{\epsilon,\gamma})$. *Then, it holds that*

$$\frac{\partial \tilde{J}_{\epsilon,\gamma}(\boldsymbol{\theta})}{\partial \theta_i} = \int_{i-1}^{i} \frac{\partial H^i(\boldsymbol{\lambda}(s|\boldsymbol{\theta}), \tilde{\boldsymbol{x}}(s|\boldsymbol{\theta}), \boldsymbol{\theta})}{\partial \theta_i} ds, \; i = 1, 2, \ldots, 2N+1, \quad (6.24)$$

*where*

$$H^i(\boldsymbol{\lambda}(s|\boldsymbol{\theta}), \tilde{\boldsymbol{x}}(s|\boldsymbol{\theta}), \boldsymbol{\theta}) = \boldsymbol{\lambda}^\top(s|\boldsymbol{\theta}) \boldsymbol{h}^i(\tilde{\boldsymbol{x}}(s|\boldsymbol{\theta}), \boldsymbol{\theta}) + \gamma \sum_{l=1}^{12} \hat{g}_{l,\epsilon}(\tilde{\boldsymbol{x}}(s|\boldsymbol{\theta})), \quad (6.25)$$

*and*

$$\boldsymbol{\lambda}(s) = (\lambda_1(s), \lambda_2(s), \lambda_3(s), \lambda_4(s), \lambda_5(s), \lambda_6(s))^\top \quad (6.26)$$

*is the solution of the costate system*

$$\dot{\boldsymbol{\lambda}}(s) = -\left(\frac{\partial H^i(\boldsymbol{\lambda}(s), \tilde{\boldsymbol{x}}(s), \boldsymbol{\theta})}{\partial \tilde{\boldsymbol{x}}}\right)^\top, \quad (6.27)$$

*with the boundary conditions*

$$\boldsymbol{\lambda}(2N+1) = (0, 0, -1, 0, 0, 0)^\top, \quad (6.28)$$

$$\boldsymbol{\lambda}(\varsigma-) = \boldsymbol{\lambda}(\varsigma+), \; \varsigma = 1, 2, \ldots, 2N. \quad (6.29)$$

*Proof.* The proof can be completed using the method of Chapter 3 in [38]. □

For each $\boldsymbol{\theta}$, the gradients of constraints in (6.14) are straightforward to be calculated. In view of these and Theorem 6.6, $(\text{PAOCP}_{\epsilon,\gamma})$ can be solved as a mathematical programming problem. As a result, we can obtain an approximately optimal solution of (AOCP) as shown in the following algorithm.

**Algorithm 6.1.**

**Step 1.** Choose initial values of $\epsilon^0$, $\gamma^0$, and $\boldsymbol{\theta}^0_{\epsilon^0,\gamma^0} \in \Theta$; set parameters $\varepsilon, \alpha, \beta, \bar{\epsilon}_1$, and $\bar{\epsilon}_2$; and set $h_1 = 0$ and $h_2 = 0$.

**Step 2.** Solve $(\text{PAOCP}_{\epsilon^{h_1},\gamma^{h_2}})$ using SQP with an initial $\boldsymbol{\theta}^0_{\epsilon^{h_1},\gamma^{h_2}}$ to an accuracy of $\varepsilon$ to give $\boldsymbol{\theta}_{\epsilon^{h_1},\gamma^{h_2}}$.

**Step 3.** Calculate $G\left(\boldsymbol{\theta}_{\epsilon^{h_1},\gamma^{h_2}}\right)$. If $G\left(\boldsymbol{\theta}_{\epsilon^{h_1},\gamma^{h_2}}\right) = 0$, then go to Step 4. Otherwise, set $\gamma^{h_2+1} := \alpha\gamma^{h_2}$ and $h_2 := h_2 + 1$. If $\gamma^{h_2} > \bar{\epsilon}_1$, then we have an abnormal exit. Otherwise, set $\boldsymbol{\theta}^0_{\epsilon^{h_1},\gamma^{h_2}} := \boldsymbol{\theta}_{\epsilon^{h_1},\gamma^{h_2-1}}$ and go to Step 2.

**Step 4.** Set $\epsilon^{h_1+1} := \beta\epsilon^{h_1}$ and $h_1 := h_1 + 1$. If $\epsilon^{h_1} > \bar{\epsilon}_2$, then set $\boldsymbol{\theta}^0_{\epsilon^{h_1},\gamma^{h_2}} := \boldsymbol{\theta}_{\epsilon^{h_1-1},\gamma^{h_2}}$ and go to Step 2. Otherwise, output $\boldsymbol{\tau}_{\epsilon^{h_1-1},\gamma^{h_2}}$ from $\boldsymbol{\theta}_{\epsilon^{h_1-1},\gamma^{h_2}}$ such that

$$\tau_{\epsilon^{h_1-1},\gamma^{h_2},i} = \sum_{k=1}^{i} \theta_{\epsilon^{h_1-1},\gamma^{h_2},k},\ i = 1,2,\ldots,2N+1, \tag{6.30}$$

and stop.

Then, $\boldsymbol{\tau}_{\epsilon^{h_1-1},\gamma^{h_2}}$ is an approximately optimal solution of (AOCP).

## 6.5 Numerical Results

In the numerical simulation, the initial condition, the number of switchings, the velocity ratio of adding alkali to glycerol, the initial concentration of glycerol in feed, the velocity of feeding glycerol, and fermentation time are $\boldsymbol{x}_0 = (0.1115\,\mathrm{g\,L^{-1}}, 495\,\mathrm{mmol\,L^{-1}}, 0, 0, 0, 5\,\mathrm{L})^\top$, $2N = 1{,}332$, $r = 0.75$, $c_{s0} = 10{,}762\,\mathrm{mmol\,L^{-1}}$, $v = 2.25873 \times 10^{-4}\,\mathrm{L\,s^{-1}}$, and $T = 85{,}788$ s, respectively.

In order to save computational time, the fermentation process is partitioned into the first batch phase (Bat. Ph.) and phases I–VIII (Phs. I–VIII). The same time durations of feeding processes (resp. batch processes) are adopted in each one of Phs. I–VIII. It should be mentioned that this approach had been adopted to obtain the experimental data in the actual fermentation process. Moreover, the bounds of the time durations in each one of Phs. I–VIII, by taking the same value for $\rho_{2\iota}$ (resp. $\delta_{2\iota}$), $\iota = 1, 2, \ldots, N$, and in Bat. Ph. are listed in Table 6.1.

Applying Algorithm 6.1 to solve the (AOCP), we obtain the optimal switching instants as shown in Table 6.2. Here, the initial values $\epsilon^0$, $\gamma^0$, and the parameters $\varepsilon, \alpha, \beta, \bar{\epsilon}_1$, and $\bar{\epsilon}_2$ are, respectively, $10^{-2}$, 1, $10^{-8}$, 2, 0.1, $10^6$, and $10^{-7}$. These parameters are derived empirically after numerous experiments. The ODEs are numerically calculated by improved Euler method with the relative error tolerance $10^{-4}$. The initial vector $\boldsymbol{\theta}^0_{\epsilon^0,\gamma^0} = \Big(\theta^0_{\epsilon^0,\gamma^0,1}, \theta^0_{\epsilon^0,\gamma^0,2}, \theta^0_{\epsilon^0,\gamma^0,3}, \ldots, \theta^0_{\epsilon^0,\gamma^0,56}, \theta^0_{\epsilon^0,\gamma^0,57}, \theta^0_{\epsilon^0,\gamma^0,58}, \theta^0_{\epsilon^0,\gamma^0,59}, \ldots, \theta^0_{\epsilon^0,\gamma^0,130}, \theta^0_{\epsilon^0,\gamma^0,131}, \theta^0_{\epsilon^0,\gamma^0,132}, \theta^0_{\epsilon^0,\gamma^0,133}, \ldots, \theta^0_{\epsilon^0,\gamma^0,252}, \theta^0_{\epsilon^0,\gamma^0,253}, \theta^0_{\epsilon^0,\gamma^0,254}, \theta^0_{\epsilon^0,\gamma^0,255}, \ldots, \theta^0_{\epsilon^0,\gamma^0,490}, \theta^0_{\epsilon^0,\gamma^0,491}, \theta^0_{\epsilon^0,\gamma^0,492}, \theta^0_{\epsilon^0,\gamma^0,493}, \ldots,$

**Table 6.1** The bounds of time durations in the Bat. Ph. and Phs. I–VIII

| Phases | Lower bounds | Upper bounds |
|---|---|---|
| Bat. Ph. | 18,720 | 19,440 |
| Phs. I–VIII | 1 | 10 |

**Table 6.2** The optimal switching instants in fed-batch fermentation

| Phases | Switching instants | Optimal values |
|---|---|---|
| Bat. Ph. | $\tau_1$ | 18,722.52 |
| Ph. I | $\tau_{2\iota}$ ($\iota = 1, \ldots, 28$) | 18,727+100.699($\iota - 1$) |
| Ph. II | $\tau_{2\iota}$ ($\iota = 29, \ldots, 65$) | 21,455.2+100.699($\iota - 29$) |
| Ph. III | $\tau_{2\iota}$ ($\iota = 66, \ldots, 126$) | 25,090.3+100.699($\iota - 66$) |
| Ph. IV | $\tau_{2\iota}$ ($\iota = 127, \ldots, 245$) | 31,142.2+100.699($\iota - 127$) |
| Ph. V | $\tau_{2\iota}$ ($\iota = 246, \ldots, 378$) | 43,031+100.699($\iota - 246$) |
| Ph. VI | $\tau_{2\iota}$ ($\iota = 379, \ldots, 459$) | 56,324.3+100.699($\iota - 379$) |
| Ph. VII | $\tau_{2\iota}$ ($\iota = 460, \ldots, 522$) | 64,381.2+100.699($\iota - 460$) |
| Ph. VIII | $\tau_{2\iota}$ ($\iota = 523, \ldots, 666$) | 70,625.5+100.699($\iota - 523$) |
| Phs. I–VIII | $\tau_{2\iota+1}$ ($\iota = 1, \ldots, 665$) | 18,722.52+100.699$\iota$ |

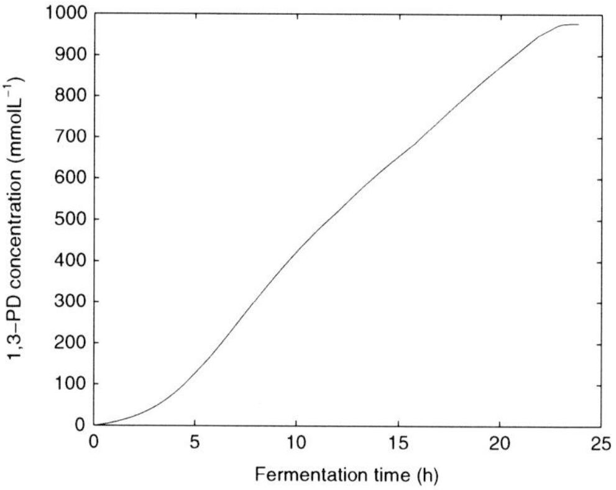


**Fig. 6.1** The change of 1,3-PD with respect to fermentation time

$\theta^0_{\epsilon^0,\gamma^0,656}$, $\theta^0_{\epsilon^0,\gamma^0,657}$, $\theta^0_{\epsilon^0,\gamma^0,658}$, $\theta^0_{\epsilon^0,\gamma^0,659}, \ldots, \theta^0_{\epsilon^0,\gamma^0,918}, \theta^0_{\epsilon^0,\gamma^0,919}, \theta^0_{\epsilon^0,\gamma^0,920}, \theta^0_{\epsilon^0,\gamma^0,921}, \ldots, \theta^0_{\epsilon^0,\gamma^0,1{,}044}, \theta^0_{\epsilon^0,\gamma^0,1{,}045}, \theta^0_{\epsilon^0,\gamma^0,1{,}046}, \theta^0_{\epsilon^0,\gamma^0,1{,}047}, \ldots, \theta^0_{\epsilon^0,\gamma^0,1{,}332}, \theta^0_{\epsilon^0,\gamma^0,1{,}333}\big)^\top$ is $(19, 188, 5, 95, \ldots, 5, 95, 7, 93, \ldots, 7, 93, 8, 92, \ldots, 8, 92, 7, 93, \ldots, 7, 93, 6, 94, \ldots, 6, 94, 4, 96, \ldots 4, 96, 3, 97, \ldots, 3, 97, 2, 98, \ldots, 2, 98)^\top$. It takes about 5 min on an AMD Athlon 64 X2 Dual Core Processor TK-57 1.90 GHz machine. Furthermore, we obtain that the concentration of 1,3-PD at the terminal time is 978.127 mmol $L^{-1}$. The concentration change of 1,3-PD obtained from the optimal switching instants is shown in Fig. 6.1.

## 6.6 Conclusion

In this chapter, we investigated the optimal control of switched autonomous system arising in constantly fed-batch fermentation. A switched autonomous system with variable switching instants was proposed to model the fermentation process.

An optimal control model involving the switched autonomous system and subject to constraint of continuous state inequality was then presented. A computational approach was developed to seek the optimal switching instants. The validity of the proposed methodology is demonstrated by numerical results.

# Chapter 7
# Optimal Control of Time-Dependent Switched Systems

## 7.1 Introduction

In this chapter, we consider optimal control of a time-dependent switched system arising in the fed-batch process. By discriminating the batch process and the feeding process, we propose a time-dependent switched system, in which the feeding rate of substrate as well as the switching instants is the control function, to mode the fed-batch process. Then, optimal control problem involving the time-dependent switched system and subject to state constraints and control constraint is presented. Especially, optimal control of such switched system has been an active research area over the past decades; see, for example, [24, 32, 186, 275]. Nevertheless, to our knowledge, the switched systems with continuous state constraints have rarely been considered. Incidentally, constrained optimal control problems have been extensively studied in the literature. Many interesting theoretical results can be found in [43]. For numerical computation, several successful families of algorithms have been developed; see, for example, [38, 92, 240, 242]. In particular, the control parameterization enhancing transform [242] has been used extensively in [133, 161, 267].

In this chapter, to seek the optimal feeding rate as well as the optimal switching instants in the constrained optimal control problem, the control parameterization enhancing transform is used to approximate the constrained optimal control problem. Then, the improved particle swarm optimization (PSO) algorithm in Chap. 5 is used to solve the resultant mathematical programming problems. Numerical results show that, by employing the obtained optimal strategy, the concentration of 1,3-PD at the terminal time can be increased considerably compared with previous results.

The main reference of this chapter is [152].



C. Liu, Z. Gong, *Optimal Control of Switched Systems Arising in Fermentation Processes*, Springer Optimization and Its Applications 97,
DOI 10.1007/978-3-662-43793-3_7

## 7.2 Time-Dependent Switched Systems

In the fed-batch process, the composition of culture medium, cultivation conditions, and analytical methods of fermentative products were similar to those previously reported [48]. Under Assumptions 5.1 and 5.2, mass balances of biomass, substrate, and products in the batch process are written as follows:

$$\begin{cases} \dot{x}_1(t) = q_1(\boldsymbol{x}(t))x_1(t), \\ \dot{x}_2(t) = -q_2(\boldsymbol{x}(t))x_1(t), \\ \dot{x}_3(t) = q_3(\boldsymbol{x}(t))x_1(t), \\ \dot{x}_4(t) = q_4(\boldsymbol{x}(t))x_1(t), \\ \dot{x}_5(t) = q_5(\boldsymbol{x}(t))x_1(t), \\ \dot{x}_6(t) = 0, \end{cases} \tag{7.1}$$

where $x_1(t), x_2(t), x_3(t), x_4(t), x_5(t)$, and $x_6(t)$ are the concentrations of biomass, glycerol, 1,3-PD, acetate, and ethanol and the volume of culture fluid at $t$ in the fermentor, respectively. $\boldsymbol{x}(t) := (x_1(t), x_2(t), x_3(t), x_4(t), x_5(t), x_6(t))^\top$ is the state vector. The specific growth rate of cells $q_1(\boldsymbol{x}(t))$, the specific consumption rate of substrate $q_2(\boldsymbol{x}(t))$, and the specific formation rates of products $q_\ell(\boldsymbol{x}(t))$, $\ell = 3, 4, 5$, are expressed by

$$q_1(\boldsymbol{x}(t)) = \frac{\Delta_1 x_2(t)}{x_2(t) + k_1} \prod_{\ell=2}^{5} \left(1 - \frac{x_\ell(t)}{x_\ell^*}\right)^{n_\ell}, \tag{7.2}$$

$$q_2(\boldsymbol{x}(t)) = m_2 + q_1(\boldsymbol{x}(t))Y_2 + \frac{\Delta_2 x_2(t)}{x_2(t) + k_2}, \tag{7.3}$$

$$q_\ell(\boldsymbol{x}(t)) = -m_\ell + q_1(\boldsymbol{x}(t))Y_\ell + \frac{\Delta_\ell x_2(t)}{x_2(t) + k_\ell}, \quad \ell = 3, 4, \tag{7.4}$$

$$q_5(\boldsymbol{x}(t)) = q_2(\boldsymbol{x}(t)) \left(\frac{c_1}{c_2 + q_1(\boldsymbol{x}(t))x_2(t)} + \frac{c_3}{c_4 + q_1(\boldsymbol{x}(t))x_2(t)}\right). \tag{7.5}$$

Under anaerobic conditions at 37 °C and pH 7.0, the kinetic parameters and critical concentrations for cell growth in (7.2)–(7.5) are as given in Table 5.1.

Due to the feed of glycerol and alkali in the fermentation process, there exist dilute effects on the concentrations of involving substances. Consequently, the batch model can be extrapolated to the feeding process by incorporating the dilution factors. Namely, mass balances of biomass, substrate, and products in the feeding process are given below:

$$\begin{cases} \dot{x}_1(t) = (q_1(\boldsymbol{x}(t)) - D(\boldsymbol{x}(t), u(t)))x_1(t), \\ \dot{x}_2(t) = D(\boldsymbol{x}(t), u(t))\left(\dfrac{c_{s0}}{1+r} - x_2(t)\right) - q_2(\boldsymbol{x}(t))x_1(t), \\ \dot{x}_3(t) = q_3(\boldsymbol{x}(t))x_1(t) - D(\boldsymbol{x}(t), u(t))x_3(t), \\ \dot{x}_4(t) = q_4(\boldsymbol{x}(t))x_1(t) - D(\boldsymbol{x}(t), u(t))x_4(t), \\ \dot{x}_5(t) = q_5(\boldsymbol{x}(t))x_1(t) - D(\boldsymbol{x}(t), u(t))x_5(t), \\ \dot{x}_6(t) = (1+r)u(t), \end{cases} \tag{7.6}$$

where $u(t) \in \mathbb{R}$ is the feeding rate of glycerol in the feed process. $c_{s0} > 0$ denotes the concentration of initial feed of glycerol in the medium. $r > 0$ is the velocity ratio of adding alkali to glycerol. Since $u(t) \geqslant 0$ and the positivity of the initial volume of culture fluid, the volume $x_6(t)$ of solution is nondecreasing and $x_6(t) > 0$. Therefore, $D(\boldsymbol{x}(t), u(t))$ is the dilution rate defined by

$$D(\boldsymbol{x}(t), u(t)) = \frac{(1+r)u(t)}{x_6(t)}. \tag{7.7}$$

Now, let $u(t)$ be the control function and suppose the switching instant $\tau_i, i \in \{1, 2, \ldots, 2N\}$, between the batch and feed processes satisfies that $0 = \tau_0 < \tau_1 < \tau_2 < \cdots < \tau_{2N} < \tau_{2N+1} = T$. In particular, $\tau_{2j+1}, j \in \Lambda_1 := \{0, 1, 2, \ldots, N\}$ is the moment of feeding glycerol, $\tau_{2j}, j \in \Lambda_2 := \{1, 2, \ldots, N\}$ is the moment of ending the feed, and $N$ is a constant in this chapter. Furthermore, denote the right-hand item of the $\ell$th equation in the system (7.1) and (7.6) by $f_\ell^i(\boldsymbol{x}(t), u(t))$ for each $\ell \in \Lambda_3 := \{1, 2, \ldots, 6\}$ and let

$$\boldsymbol{f}^i(\boldsymbol{x}(t), u(t)) = \left(f_1^i(\boldsymbol{x}(t), u(t)), \ldots, f_6^i(\boldsymbol{x}(t), u(t))\right)^\top. \tag{7.8}$$

Then, the nonlinear time-dependent switched system describing the whole process of fed-batch fermentation can be formulated as

$$\begin{cases} \dot{\boldsymbol{x}}(t) = \boldsymbol{f}^i(\boldsymbol{x}(t), u(t)), \\ u(t) \in U_i, \ t \in (\tau_{i-1}, \tau_i], \\ \boldsymbol{x}(\tau_{i-1}+) = \boldsymbol{x}(\tau_{i-1}), \ i = 1, 2, \ldots, 2N+1, \\ \boldsymbol{x}(0) = \boldsymbol{x}_0. \end{cases} \tag{7.9}$$

where $\boldsymbol{x}_0$ is a given initial state, the notation + indicates the limit from the right, and

$$U_i = \begin{cases} [a_i, b_i], & \text{if } i \text{ is even}, \\ \{0\}, & \text{if } i \text{ is odd}. \end{cases} \tag{7.10}$$

Here, $a_{2j-1}$ and $b_{2j-1}$, $j \in \Lambda_2$, are positive constants which denote the minimal and maximal rates of adding glycerol, respectively. Thus, we define the class of admissible control functions as

$$\mathcal{U} := \{u \in L_\infty([0,T],\mathbb{R}) |\ u(t) \in U_i, i = 1,2,\ldots,2N+1\}, \tag{7.11}$$

where $L_\infty([0,T],\mathbb{R})$ is the Banach space of all essentially bounded functions from $[0,T]$ into $\mathbb{R}$.

Since biological considerations limit the rate of switching, there are maximal and minimal time durations that are spent on each of the batch and feeding processes. On this basis, define the set of admissible switching instants as

$$\Gamma := \{(\tau_1, \tau_2, \ldots, \tau_{2N})^\top \in \mathbb{R}^{2N} |\ \rho_i \leqslant \tau_i - \tau_{i-1} \leqslant \delta_i, \quad i \in \Lambda := \{1,2,\ldots,2N+1\}\}, \tag{7.12}$$

where $\rho_i$ and $\delta_i$ are the minimal and maximal time durations, respectively. Accordingly, any $\boldsymbol{\tau} \in \Gamma$ is regarded as an admissible vector of switching instants.

There exist critical concentrations of biomass, glycerol, 1,3-PD, acetate, and ethanol, outside which cells cease to grow. Hence, it is biologically meaningful to restrict the concentrations of biomass, glycerol, and products and the volume of culture fluid within a set $\tilde{W}$ defined as

$$\boldsymbol{x}(t) \in \tilde{W} := \prod_{\ell=1}^{6} \left[x_{*\ell}, x_\ell^*\right], \ \forall t \in [0,T], \tag{7.13}$$

where $x_{*\ell}$, $x_\ell^*$, $\ell = 1,2,\ldots,5$, are as given in Table 5.1, $x_{*6} = 4$ and $x_6^* = 7$.

For the system (7.9), some important properties are discussed as follows.

**Theorem 7.1.** *The functions* $\boldsymbol{f}^i(\cdot,\cdot)$, $i = 1,2,\ldots,2N+1$, *in the system* (7.9) *satisfy the following conditions:*

(a) $\boldsymbol{f}^i(\cdot,\cdot)$ *are affine in control* $u$,
(b) $\boldsymbol{f}^i(\cdot,\cdot) : \mathbb{R}_+^6 \times U_i \to \mathbb{R}^6$, *together with its partial derivatives with respect to* $\boldsymbol{x}$ *and* $u$, *are continuous on* $\mathbb{R}_+^6 \times U_i$.
(c) *There exists a constant* $K > 0$ *such that*

$$\|\boldsymbol{f}^i(\boldsymbol{x},u)\| \leqslant K(1 + \|\boldsymbol{x}\|), \forall (\boldsymbol{x},u) \in \mathbb{R}_+^6 \times U_i, \tag{7.14}$$

*where* $\|\cdot\|$ *is the Euclidean norm.*

*Proof.* (a) It is easy to verify that $\boldsymbol{f}^i$ is affine in control $u$ by its definition.
(b) This conclusion can be obtained by the expressions of $\boldsymbol{f}^i$ in (7.1) and (7.6).
(c) We can complete the proof using a method similar to the proof of Proposition 5.1 in Chap. 5. □

**Theorem 7.2.** *For each $u \in \mathcal{U}$ and $\boldsymbol{\tau} \in \Gamma$, the system (7.9) has a unique continuous solution denoted by $\boldsymbol{x}(\cdot|u, \boldsymbol{\tau})$. Furthermore, $\boldsymbol{x}(\cdot|u, \boldsymbol{\tau})$ satisfies the following integral equation*

$$\boldsymbol{x}(t|u, \boldsymbol{\tau}) = \boldsymbol{x}(\tau_{i-1}|u) + \int_{\tau_{i-1}}^{t} \boldsymbol{f}^i(\boldsymbol{x}(s|u, \boldsymbol{\tau}), u(s))\mathrm{d}s,$$
$$\forall t \in (\tau_{i-1}, \tau_i],\ i = 1, 2, \ldots, 2N+1, \tag{7.15}$$

*and is continuous in $u$ and $\boldsymbol{\tau}$.*

*Proof.* This conclusion can be obtained from Theorem 7.1 and the theory of ordinary differential equations [5]. □

**Theorem 7.3.** *If $\boldsymbol{x}(\cdot|u, \boldsymbol{\tau})$ is a solution of the system (7.9) with given initial condition $\boldsymbol{x}_0$, then it is uniformly bounded.*

*Proof.* In view of Theorems 7.1 and 7.2, we obtain that, for each $u \in \mathcal{U}$ and $\boldsymbol{\tau} \in \Gamma$,

$$\begin{aligned}\|\boldsymbol{x}(t|u, \boldsymbol{\tau})\| \leqslant\ & \|\boldsymbol{x}_0\| + \sum_{j=1}^{i-1} \int_{\tau_{j-1}}^{\tau_j} \|\boldsymbol{f}^j(\boldsymbol{x}(s|u, \boldsymbol{\tau}), u(s))\|\mathrm{d}s \\ & + \int_{\tau_{i-1}}^{t} \|\boldsymbol{f}^i(\boldsymbol{x}(s|u, \boldsymbol{\tau}), u(s))\|\mathrm{d}s \\ \leqslant\ & \|\boldsymbol{x}_0\| + K\int_0^t (1 + \|\boldsymbol{x}(s|u, \boldsymbol{\tau})\|)\mathrm{d}s, \\ & \forall t \in (\tau_{i-1}, \tau_i],\ i = 1, 2, \ldots, 2N+1.\end{aligned}$$

By Lemma 4.1, it follows that

$$\|\boldsymbol{x}(t|u, \boldsymbol{\tau})\| \leqslant M_1', \quad \forall t \in [0, T],$$

where $M_1' := (\|\boldsymbol{x}_0\| + KT)\exp(KT)$. □

## 7.3 Constrained Optimal Control Problems

Basically, the control task of fed-batch fermentation lies in the determination of the proper feeding rate of glycerol and the switching instants between the batch and feeding processes to obtain as much 1,3-PD as possible at the terminal time of the fermentation. Especially, physical limitations (7.11)–(7.13) are set during the fermentation process.

Then, the constrained optimal control problem may be stated formally as **Problem (P)**. Given the switched system

$$\begin{cases} \dot{\boldsymbol{x}}(t) = \boldsymbol{f}^i(\boldsymbol{x}(t), u(t)), \\ u(t) \in U_i, \ t \in (\tau_{i-1}, \tau_i], \\ \boldsymbol{x}(\tau_{i-1}+) = \boldsymbol{x}(\tau_{i-1}), \ i = 1, 2, \ldots, 2N+1, \\ \boldsymbol{x}(0) = \boldsymbol{x}_0, \end{cases}$$

find a control $u \in \mathcal{U}$ and a switching vector $\boldsymbol{\tau} \in \Gamma$ such that the constraint (7.13) is satisfied and the cost functional

$$J(u, \boldsymbol{\tau}) = -x_3(T|u, \boldsymbol{\tau}) \tag{7.16}$$

is minimized.

By similar arguments as those given for Theorem 5.3 in Chap. 5, we confirm the existence of the optimal solution for Problem (P).

**Theorem 7.4.** *Problem* (P) *has at least one optimal solution.*

## 7.4 Computational Approaches

In this section, we shall develop a numerical solution method to Problem (P).

### 7.4.1 Approximate Problem

For each $p_i \geqslant 1, i \in \{1, 2, \ldots, 2N+1\}$, let the time subinterval $[\tau_{i-1}, \tau_i]$ be partitioned into $n_{p_i}$ subintervals with $n_{p_i} + 1$ partition points denoted by

$$\tau_0^{p_i}, \tau_1^{p_i}, \ldots, \tau_{n_{p_i}}^{p_i}, \tau_0^{p_i} = \tau_{i-1}, \tau_{n_{p_i}}^{p_i} = \tau_i, \text{ and } \tau_{k-1}^{p_i} \leqslant \tau_k^{p_i}.$$

Let $n_{p_i}$ be chosen such that $n_{p_i+1} \geqslant n_{p_i}$. The control is now approximated in the form of a piecewise constant function as follows:

$$u^p(t) = \sum_{i=1}^{2N+1} \sum_{k=1}^{n_{p_i}} \sigma^{p_i,k} \chi_{(\tau_{k-1}^{p_i}, \tau_k^{p_i}]}(t). \tag{7.17}$$

Here, $\chi_{(\tau_{k-1}^{p_i}, \tau_k^{p_i}]}$ is the indicator function on the interval $\left(\tau_{k-1}^{p_i}, \tau_k^{p_i}\right]$ defined by

$$\chi_{(\tau_{k-1}^{p_i}, \tau_k^{p_i}]}(t) = \begin{cases} 1, \ t \in \left(\tau_{k-1}^{p_i}, \tau_k^{p_i}\right], \\ 0, \text{ otherwise.} \end{cases}$$

Let $\kappa := \sum_{i=1}^{2N+1} n_{p_i}$. Then, $\boldsymbol{\sigma}^p := \left((\boldsymbol{\sigma}^{p_1})^\top, \ldots, (\boldsymbol{\sigma}^{p_{2N+1}})^\top\right)^\top \in \mathbb{R}^\kappa$, where $\boldsymbol{\sigma}^{p_i} := (\sigma^{p_i,1}, \ldots, \sigma^{p_i,n_{p_i}})^\top$ defines the heights of the approximate control (7.17). From (7.11), it is clear that

$$\sigma^{p_i,k} \in U_i,\ k = 1,2,\ldots,n_{p_i},\ i = 1,2,\ldots,2N+1. \tag{7.18}$$

Let $\Xi^p$ be the set of all those $\boldsymbol{\sigma}^p$ which satisfy the constraints (7.18).

Note that $\tau_k^{p_i}$, $k = 1,2,\ldots,n_{p_i}, i = 1,2,\ldots,2N+1$, taken as the decision variables will encounter numerical difficulties as mentioned in [240]. For this reason, a control parameterization enhancing transform is introduced to map these variable time points into preassigned fixed knots in a new time scale. It is achieved by introducing a transform from $t \in [0,T]$ to $s \in [0, 2N+1]$ as follows:

$$\dot{t}(s) = v^p(s), \quad t(0) = 0, \tag{7.19}$$

where $v^p$ is given by

$$v^p(s) = \sum_{i=1}^{2N+1}\sum_{k=1}^{n_{p_i}} \theta_k^{p_i} \chi_{\left(i-1+\frac{k-1}{n_{p_i}},\, i-1+\frac{k}{n_{p_i}}\right]}(s). \tag{7.20}$$

In (7.20), $\theta_k^{p_i} \geqslant 0, k = 1,2,\ldots,n_{p_i}, i = 1,2,\ldots,2N+1$, are decision variables. Let $\boldsymbol{\theta}^p$ be the vector whose components are $\theta_k^{p_i}, k = 1,2,\ldots,n_{p_i}, i = 1,2,\ldots,2N+1$, and $\Omega^p$ be the set of all such $\boldsymbol{\theta}^p$. Clearly,

$$t(s) = \int_0^s v^p(\eta)\mathrm{d}\eta = \sum_{l=1}^{i-1}\sum_{j=1}^{n_{p_l}} \frac{\theta_j^{p_l}}{n_{p_i}} + \sum_{j=1}^{k-1}\theta_j^{p_i} + \theta_k^{p_i}\left(s - i + 1 - \frac{k-1}{n_{p_i}}\right),$$

$$\forall s \in \left(i-1+\frac{k-1}{n_{p_i}},\, i-1+\frac{k}{n_{p_i}}\right], \tag{7.21}$$

and

$$t(2N+1) = \sum_{i=1}^{2N+1}\sum_{k=1}^{n_{p_i}} \theta_k^{p_i} = T. \tag{7.22}$$

Let

$$w^p(s) = u^p(t(s)).$$

Then

$$w^p(s) = \sum_{i=1}^{2N+1}\sum_{k=1}^{n_{p_i}} \sigma^{p_i,k} \chi_{\left(i-1+\frac{k-1}{n_{p_i}},\, i-1+\frac{k}{n_{p_i}}\right]}(s).$$

Define

$$\tilde{\boldsymbol{x}}(s) := \left(\boldsymbol{x}(s)^{\top}, t(s)\right)^{\top}$$

and

$$\tilde{\boldsymbol{f}}^{i}(\tilde{\boldsymbol{x}}(s), \boldsymbol{\sigma}^{p}, \boldsymbol{\theta}^{p}) := (v^{p}(s)\boldsymbol{f}^{i}(\boldsymbol{x}(t(s)), w^{p}(s|\boldsymbol{\sigma}^{p}))^{\top}, v^{p}(s|\boldsymbol{\theta}^{p}))^{\top}.$$

Let $\tilde{\boldsymbol{x}}(\cdot|\boldsymbol{\sigma}^{p}, \boldsymbol{\theta}^{p})$ be the solution of the following system corresponding to the control parameter vector $(\boldsymbol{\sigma}^{p}, \boldsymbol{\theta}^{p}) \in \Xi^{p} \times \Omega^{p}$:

$$\begin{cases} \dot{\tilde{\boldsymbol{x}}}(s) = \tilde{\boldsymbol{f}}^{i}(\tilde{\boldsymbol{x}}(s), \boldsymbol{\sigma}^{p}, \boldsymbol{\theta}^{p}), \ s \in (i-1, i], \\ \tilde{\boldsymbol{x}}(i-1+) = \tilde{\boldsymbol{x}}(i-1), \ i = 1, 2, \ldots, 2N+1, \\ \tilde{\boldsymbol{x}}(0) = (\boldsymbol{x}_0^{\top}, 0)^{\top}. \end{cases}$$

Then, the constraint (7.13) can be rewritten as

$$(\tilde{x}_1(s), \tilde{x}_2(s), \tilde{x}_3(s), \tilde{x}_4(s), \tilde{x}_5(s), \tilde{x}_6(s))^{\top} \in \tilde{W}, \ \forall s \in [0, 2N+1]. \tag{7.23}$$

Now, we may specify the approximate problem (P(p)) as follows:

**Problem (P(p))**. Subject to the system

$$\begin{cases} \dot{\tilde{\boldsymbol{x}}}(t) = \tilde{\boldsymbol{f}}^{i}(\tilde{\boldsymbol{x}}(s), \boldsymbol{\sigma}^{p}, \boldsymbol{\theta}^{p}), \ s \in (i-1, i], \\ \tilde{\boldsymbol{x}}(i-1+) = \tilde{\boldsymbol{x}}(i-1), \ i = 1, 2, \ldots, 2N+1, \\ \tilde{\boldsymbol{x}}(0) = (\boldsymbol{x}_0^{\top}, 0)^{\top}, \end{cases}$$

find a combined vector $(\boldsymbol{\sigma}^{p}, \boldsymbol{\theta}^{p}) \in \Xi^{p} \times \Omega^{p}$ such that the constraints (7.22) and (7.23) are satisfied and the cost functional

$$J(\boldsymbol{\sigma}^{p}, \boldsymbol{\theta}^{p}) = -\tilde{x}_3(2N+1|\boldsymbol{\sigma}^{p}, \boldsymbol{\theta}^{p}) \tag{7.24}$$

is minimized.

### *7.4.2 Continuous State Constraints*

Since constraint (7.23) in Problem (P(p)) is a continuous state constraint, we shall apply the method in Chap. 6 to deal with this constraint.

Let

$$g_{\ell}(\tilde{\boldsymbol{x}}(s|\boldsymbol{\sigma}^{p}, \boldsymbol{\theta}^{p})) = \tilde{x}_{\ell}(s|\boldsymbol{\sigma}^{p}, \boldsymbol{\theta}^{p}) - \tilde{x}_{\ell}^{*},$$

$$g_{6+\ell}(\tilde{\boldsymbol{x}}(s|\boldsymbol{\sigma}^{p}, \boldsymbol{\theta}^{p})) = \tilde{x}_{*\ell} - \tilde{x}_{\ell}(s|\boldsymbol{\sigma}^{p}, \boldsymbol{\theta}^{p}), \ \ell = 1, 2, \ldots, 6.$$

The constraint (7.23) is equivalently transcribed into

$$G(\boldsymbol{\sigma}^p, \boldsymbol{\theta}^p) = 0, \tag{7.25}$$

where $G(\boldsymbol{\sigma}^p, \boldsymbol{\theta}^p) := \sum_{l=1}^{12} \int_0^{2N+1} \max\{0, g_l(\tilde{\boldsymbol{x}}(s|\boldsymbol{\sigma}^p, \boldsymbol{\theta}^p))\}\mathrm{d}s$. However, $G(\cdot, \cdot)$ is non-differentiable at the point $g_l = 0, l \in \{1, 2, \ldots, 12\}$. We replace (7.25) with

$$\tilde{G}_{\varepsilon,\gamma}(\boldsymbol{\sigma}^p, \boldsymbol{\theta}^p) := \gamma + \sum_{l=1}^{12} \int_0^{2N+1} \varphi_\varepsilon(g_l(\tilde{\boldsymbol{x}}(s|\boldsymbol{\sigma}^p, \boldsymbol{\theta}^p)))\mathrm{d}s \geqslant 0, \tag{7.26}$$

where $\varepsilon > 0$, $\gamma > 0$, and

$$\varphi_\varepsilon(\eta) = \begin{cases} \eta, & \text{if } \eta < -\varepsilon, \\ \dfrac{(\eta+\varepsilon)^2}{4\varepsilon}, & \text{if } -\varepsilon \leqslant \eta \leqslant \varepsilon, \\ 0, & \text{if } \eta > \varepsilon. \end{cases} \tag{7.27}$$

Thus, Problem (P(p)) is approximated by a sequence of Problems $\{(\mathrm{P}_{\varepsilon,\gamma}(\mathrm{p}))\}$ defined by replacing constraint (7.25) with (7.26). Then, the gradient of the constraint $\tilde{G}_{\varepsilon,\gamma}(\cdot, \cdot)$ can be computed by the following theorem.

**Theorem 7.5.** *For the constraint $\tilde{G}_{\varepsilon,\gamma}(\boldsymbol{\sigma}^p, \boldsymbol{\theta}^p)$ given in (7.26), it holds that its gradients with respect to parameterized control $\boldsymbol{\sigma}^p$ and $\boldsymbol{\theta}^p$ are, respectively,*

$$\frac{\partial \tilde{G}_{\varepsilon,\gamma}(\boldsymbol{\sigma}^p, \boldsymbol{\theta}^p)}{\partial \sigma^{p_i,k}} = \int_{i-1+\frac{k-1}{n_{p_i}}}^{i-1+\frac{k}{n_{p_i}}} \frac{\partial \tilde{H}(\tilde{\boldsymbol{x}}(s|\boldsymbol{\sigma}^p, \boldsymbol{\theta}^p), \boldsymbol{\sigma}^p, \boldsymbol{\theta}^p, \tilde{\boldsymbol{\lambda}}(s))}{\partial \sigma^{p_i,k}} \mathrm{d}s,$$

*and*

$$\frac{\partial \tilde{G}_{\varepsilon,\gamma}(\boldsymbol{\sigma}^p, \boldsymbol{\theta}^p)}{\partial \theta^{p_i,k}} = \int_{i-1+\frac{k-1}{n_{p_i}}}^{i-1+\frac{k}{n_{p_i}}} \frac{\partial \tilde{H}(\tilde{\boldsymbol{x}}(s|\boldsymbol{\sigma}^p, \boldsymbol{\theta}^p), \boldsymbol{\sigma}^p, \boldsymbol{\theta}^p, \tilde{\boldsymbol{\lambda}}(s))}{\partial \theta^{p_i,k}} \mathrm{d}s,$$

*where*

$$\begin{aligned}\tilde{H}(\tilde{\boldsymbol{x}}(s|\boldsymbol{\sigma}^p, \boldsymbol{\theta}^p), \boldsymbol{\sigma}^p, \boldsymbol{\theta}^p, \tilde{\boldsymbol{\lambda}}(s)) &= \sum_{l=1}^{12} \varphi_\varepsilon(g_l(\tilde{\boldsymbol{x}}(s|\boldsymbol{\sigma}^p, \boldsymbol{\theta}^p))) \\ &\quad + \tilde{\boldsymbol{\lambda}}^\top(s) \tilde{\boldsymbol{f}}^i(\tilde{\boldsymbol{x}}(s|\boldsymbol{\sigma}^p, \boldsymbol{\theta}^p), \boldsymbol{\sigma}^p, \boldsymbol{\theta}^p),\end{aligned}$$

*and*

$$\tilde{\boldsymbol{\lambda}}(s) = \left(\tilde{\lambda}_1(s), \tilde{\lambda}_2(s), \tilde{\lambda}_3(s), \tilde{\lambda}_4(s), \tilde{\lambda}_5(s), \tilde{\lambda}_6(s), \tilde{\lambda}_7(s)\right)^\top$$

*is the solution of the costate system*

$$\dot{\tilde{\boldsymbol{\lambda}}}(s) = -\left(\frac{\partial \tilde{H}\left(\tilde{\boldsymbol{x}}(s|\boldsymbol{\sigma}^p, \boldsymbol{\theta}^p), \boldsymbol{\sigma}^p, \boldsymbol{\theta}^p, \tilde{\boldsymbol{\lambda}}(s)\right)}{\partial \tilde{\boldsymbol{x}}}\right)^\top,$$

*with the boundary conditions*

$$\tilde{\boldsymbol{\lambda}}(2N+1) = (0,0,0,0,0,0,0)^\top,$$
$$\tilde{\boldsymbol{\lambda}}(0) = (0,0,0,0,0,0,0)^\top,$$
$$\tilde{\boldsymbol{\lambda}}(i-) = \tilde{\boldsymbol{\lambda}}(i+), \quad i = 1,2,\ldots,2N.$$

*Proof.* We can complete the proof using a method similar to the proof of Theorem 5.3. in [38]. □

### 7.4.3 Optimization Algorithms

Each of Problems $\{(\mathrm{P}_{\varepsilon,\gamma}(\mathrm{p}))\}$ is in essence a mathematical programming problem which can be solved by various optimization methods such as gradient-based techniques [38, 240]. However, all those techniques are only designed to find local optimal solutions. In contrast, stochastic evolution methods generally lead to better results than deterministic ones. An improved PSO algorithm was developed to solve the optimal control problems with control and state constraints, and its effectiveness was demonstrated in Chap. 5. Therefore, we will solve each Problem $(P_{\varepsilon,\gamma}(\mathrm{p}))$ using the improved PSO algorithm as in Chap. 5.

Based on the improved PSO algorithm in Chap. 5, we can obtain an approximately optimal control and optimal switching instants for Problem (P) as shown in the following algorithm.

**Algorithm 7.1.**

**Step 1.** Choose initial values of $\varepsilon$ and $\gamma$; set parameters $\alpha > 0$ and $\beta > 0$.
**Step 2.** Solve Problem $(\mathrm{P}_{\varepsilon,\gamma}(\mathrm{p}))$ using the improved PSO algorithm to give $\left(\boldsymbol{\sigma}^{p,*}_{\varepsilon,\gamma}, \boldsymbol{\delta}^{p,*}_{\varepsilon,\gamma}\right)$.

**Step 3.** Check feasibility of $g_l\left(\tilde{\boldsymbol{x}}\left(s|\boldsymbol{\sigma}^{p,*}_{\varepsilon,\gamma}, \boldsymbol{\delta}^{p,*}_{\varepsilon,\gamma}\right)\right) \geq 0$ for $s \in [0, 2N+1]$, $l = 1,2,\ldots,12$.

**Step 4.** If $\left(\boldsymbol{\sigma}_{\varepsilon,\gamma}^{p,*}, \boldsymbol{\delta}_{\varepsilon,\gamma}^{p,*}\right)$ is feasible, then go to Step 5. Otherwise, set $\gamma := \alpha\gamma$. If $\gamma < \bar{\gamma}$, where $\bar{\gamma}$ is a prespecified positive constant, then go to Step 6. Otherwise go to Step 2.

**Step 5.** Set $\varepsilon := \beta\varepsilon$. If $\varepsilon > \bar{\varepsilon}$, where $\bar{\varepsilon}$ is a prespecified positive constant, then go to Step 3. Otherwise go to Step 6.

**Step 6.** If $\min_{i\in\{1,2,\ldots,2N+1\}} n_{p_i} \geqslant P$, where $P$ is a predefined positive constant, then go to Step 7. Otherwise, go to Step 1 with $n_{p_i}$ increased to $n_{p_i+1}$ for each $i$.

**Step 7.** Stop and construct $u^{p,*}$ and $\boldsymbol{\tau}^{p,*}$ from $\left(\boldsymbol{\sigma}_{\varepsilon,\gamma}^{p,*}, \boldsymbol{\delta}_{\varepsilon,\gamma}^{p,*}\right)$ according to (7.17) and (7.21).

Then, $(u^{p,*}, \boldsymbol{\tau}^{p,*})$ obtained is an approximately optimal solution of Problem (P).

*Remark 7.1.* In the algorithm, $\varepsilon$ is a parameter controlling the accuracy of the smoothing approximation. $\gamma$ is a parameter controlling the feasibility of the constraint (7.23).

*Remark 7.2.* It is important for the validity of the above algorithm to choose the parameters $\alpha$, $\beta$, $\bar{\varepsilon}$, and $\bar{\gamma}$. Especially, the parameters $\alpha$ and $\beta$ must be chosen less than 1. $\bar{\varepsilon}$ and $\bar{\gamma}$ are two sufficient small values such that the algorithm is effective.

## 7.5 Numerical Results

In the numerical simulation, to solve numerically the nonlinear time-dependent switched system (7.9), the initial state, velocity ratio of adding alkali to glycerol, concentration of initial feed glycerol, fermentation time, and the number of switchings are $\boldsymbol{x}_0 = (0.1115\,\mathrm{g\,L^{-1}}, 495\,\mathrm{mmol\,L^{-1}}, 0, 0, 0, 5\,\mathrm{L})^\top$, $r = 0.75$, $c_{s0} = 10{,}762\,\mathrm{mmol\,L^{-1}}$, $T = 24.16\,\mathrm{h}$, and $2N = 1{,}354$, respectively.

In order to save computational time, the fermentation process is partitioned into the first batch phase (Bat. Ph.) and phases I–IX (Phs. I–IX) according to the number of switchings. The same feed strategies are adopted in each one of Phs. I–IX. Furthermore, the time durations for two adjacent processes, i.e., a feeding process and its succeeding batch process, in Phs. I–IX are equal and assumed to be $\dfrac{3{,}600 \times (T - \tau_1)}{N}$ seconds. It should be mentioned that this approach had been adopted to calculate the computational results in Chap. 5 and to obtain the experimental data in the fermentation process. Moreover, the bounds of feeding rates in Phs. I–IX are chosen as in Table 5.2. The bounds of the time duration in each of Phs. I–IX are listed in Table 7.1.

In the improved PSO algorithm, the number of initial particle swarm $N^p$, the maximal iteration $M^p$, and the parameters $P_{cr}$, $M_1^p$, $M_2^p$, $c_1$, $c_2$, and $\varepsilon_p$ are, respectively, 200, 100, 0.5, 50,10, 2, 2, and $10^{-3}$. In Algorithm 7.1, the initial values of $u$ and $\boldsymbol{\tau}$ are chosen as those in Chap. 5, and the smoothing and feasible parameters were initially selected as $\varepsilon = 0.1$ and $\gamma = 0.01$ and then subsequently adjusted according to the guidelines in Algorithm 7.1. In particular, the parameters $\alpha$ and $\beta$

**Table 7.1** The bounds of time durations in the Bat. Ph. and Phs. I–IX

| Phases | Bounds | Values (s) | Bounds | Values (s) |
|---|---|---|---|---|
| Bat. Ph. | $\rho_1$ | 19,080 | $\delta_1$ | 19,440 |
| Ph. I | $\rho_{2j}$ | 2 | $\delta_{2j}$ | 8 |
| $(j = 1, \ldots, 28)$ | $\rho_{2j+1}$ | 92 | $\delta_{2j+1}$ | 98 |
| Phs. II–V | $\rho_{2j}$ | 4 | $\delta_{2j}$ | 10 |
| $(j = 29, \ldots, 378)$ | $\rho_{2j+1}$ | 90 | $\delta_{2j+1}$ | 96 |
| Phs. VI–VIII | $\rho_{2j}$ | 1 | $\delta_{2j}$ | 7 |
| $(j = 379, \ldots, 666)$ | $\rho_{2j+1}$ | 93 | $\delta_{2j+1}$ | 99 |
| Ph. IX | $\rho_{2j}$ | 1 | $\delta_{2j}$ | 3 |
| $(j = 667, \ldots, 677)$ | $\rho_{2j+1}$ | 97 | $\delta_{2j+1}$ | 99 |

**Table 7.2** The optimal switching instants in the fed-batch fermentation process

| Phases | Switching instants | Optimal values (s) |
|---|---|---|
| Bat. Ph. | $\tau_1$ | 19,084.9 |
| Ph. I | $\tau_{2j}$ | 19,084.9+100.282$(j-1)$ |
| $(j = 1, \ldots, 28)$ | $\tau_{2j+1}$ | $19,092.543 + 100.282j$ |
| Ph. II | $\tau_{2j}$ | 21,901.369+100.282$(j-29)$ |
| $(j = 29, \ldots, 65)$ | $\tau_{2j+1}$ | 21,892.8+100.282$(j-28)$ |
| Ph. III | $\tau_{2j}$ | 25,613.049+100.282$(j-66)$ |
| $(j = 66, \ldots, 126)$ | $\tau_{2j+1}$ | 25,603.2+100.282$(j-65)$ |
| Ph. IV | $\tau_{2j}$ | 31,730.356+100.282$(j-127)$ |
| $(j = 127, \ldots, 245)$ | $\tau_{2j+1}$ | 31,720.4+100.282$(j-126)$ |
| Ph. V | $\tau_{2j}$ | 43,660.404+100.282$(j-246)$ |
| $(j = 246, \ldots, 378)$ | $\tau_{2j+1}$ | 43,654+100.282$(j-245)$ |
| Ph. VI | $\tau_{2j}$ | 56,992.716 +100.282$(j-379)$ |
| $(j = 379, \ldots, 459)$ | $\tau_{2j+1}$ | 56,991.6+100.282$(j-378)$ |
| Ph. VII | $\tau_{2j}$ | 65,115.519+100.282$(j-460)$ |
| $(j = 460, \ldots, 522)$ | $\tau_{2j+1}$ | 65,114.5+100.282$(j-459)$ |
| Ph. VIII | $\tau_{2j}$ | 71,433.213+100.282$(j-523)$ |
| $(j = 523, \ldots, 666)$ | $\tau_{2j+1}$ | 71,432.2+100.282$(j-522)$ |
| Ph. IX | $\tau_{2j}$ | 85,873.987+100.282$(j-667)$ |
| $(j = 667, \ldots, 677)$ | $\tau_{2j+1} (j \neq 677)$ | 85,872.9+100.282$(j-666)$ |

were chosen as 0.1 and 0.01 until the solution obtained is feasible for the original problem. The process was terminated when $\bar{\varepsilon} = 1.0 \times 10^{-8}$ and $\bar{\gamma} = 1.0 \times 10^{-7}$. It is worth mentioning that in the first step, a small value of $\gamma$ was required to ensure feasibility. After that the $\gamma$ hardly changed as $\varepsilon$ was decreased. The specified constant $P$ in Algorithm 7.1 is 2. Note also that only a small improvement (less than 0.01) was obtained by resolving the problem with 5.

Applying Algorithm 7.1 to the Problem (P), we obtain the optimal feeding rates of glycerol in Phs. I–IX as shown in Fig. 7.1 and the optimal switching instants listed in Table 7.2. Here, all the computations are performed in Visual C++ 6.0,

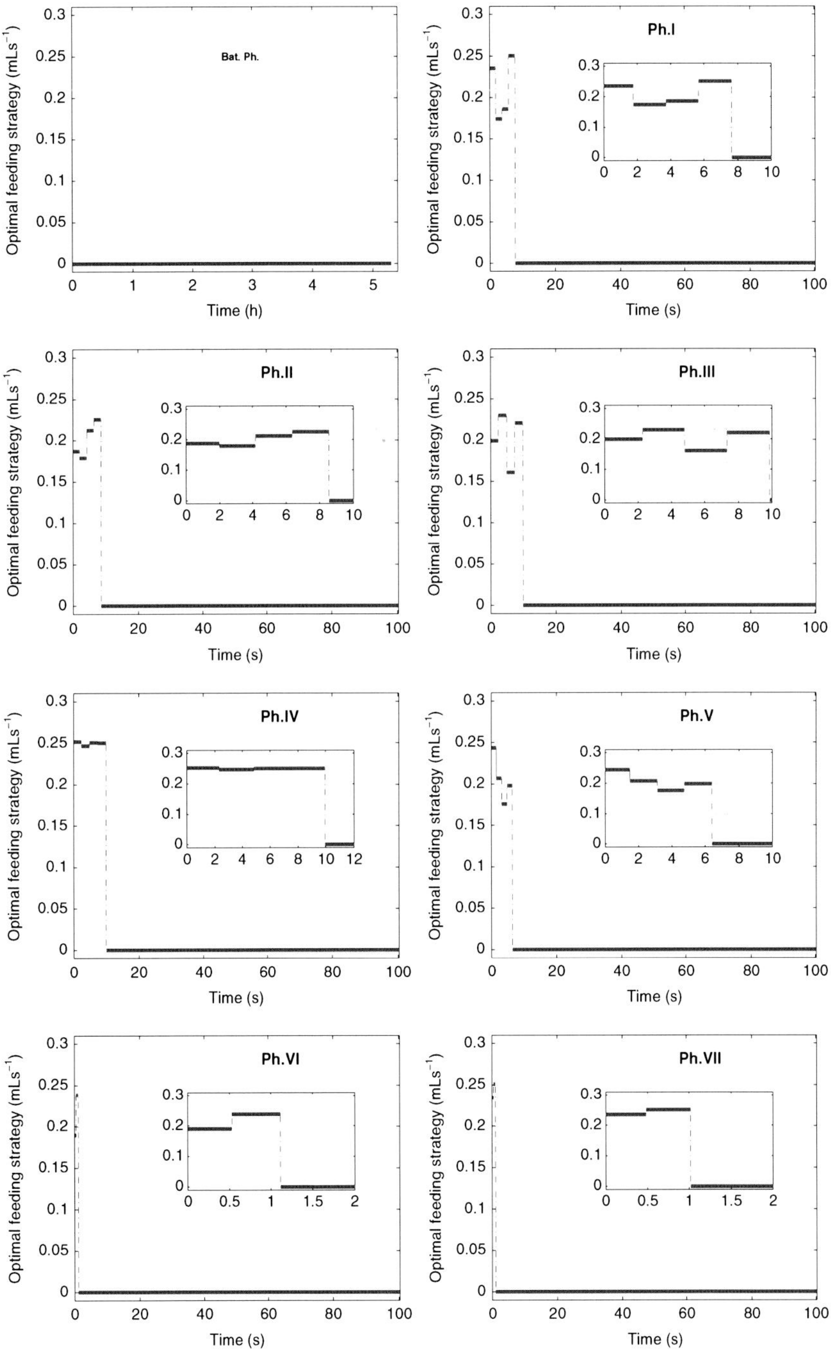


**Fig. 7.1** The optimal feeding strategy of glycerol in the fed-batch fermentation process

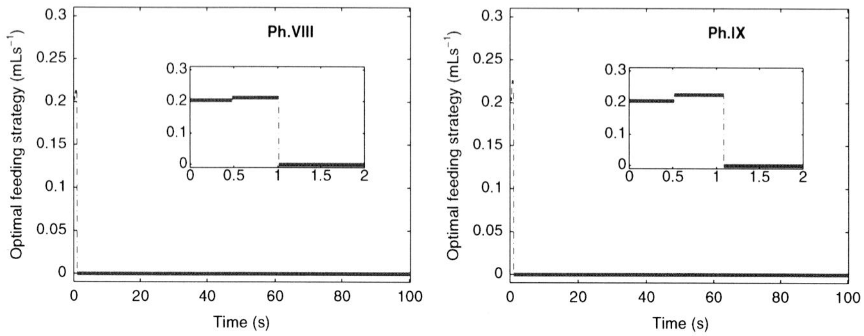


**Fig. 7.1** (continued)

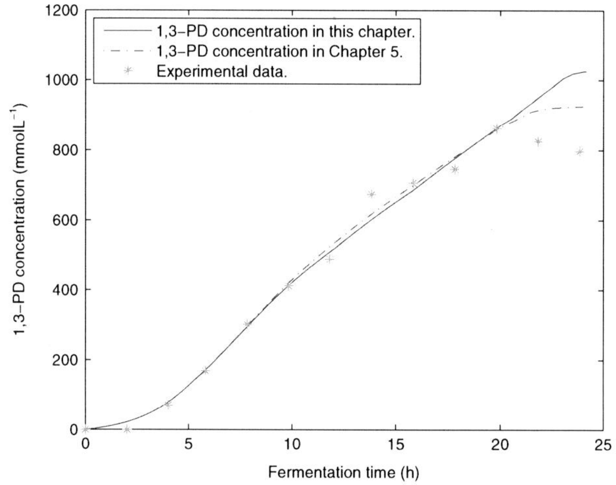


**Fig. 7.2** The changes of 1,3-PD concentration with respect to time in the fed-batch fermentation process

and numerical results are plotted by Matlab 7.10.0. In particular, the ODEs in the computation process are numerically calculated by improved Euler method [94] with the relative error tolerance $10^{-4}$. In detail, the line in the first subfigure of Fig. 7.1 indicates the feeding rate of glycerol, which is identically equal to zero, and the time duration in the Bat. Ph. Accordingly, the lines in the next nine subfigures illustrate the feeding rates of glycerol in conjunction with time durations of a feeding process and its succeeding batch process in Phs. I–IX, respectively. To show the feeding rates of glycerol in the feeding processes for Phs. I–IX better, nine small subfigures are also incorporated in the corresponding nine subfigures, respectively.

Under the obtained optimal feeding rates and the optimal switching instants, the computational concentration of 1,3-PD at the terminal time is $1025.3\,\text{mmol}\,\text{L}^{-1}$ which is increased by $28.64\,\%$ in comparison with experimental result

797.23 mmol $L^{-1}$. Furthermore, compared with the obtained 1,3-PD concentration 925.127 mmol $L^{-1}$ in Chap. 5, which is computed in case that the number of phases is the same and the switching instants between the batch and feeding processes are decided a priori, the concentration of 1,3-PD at the terminal time obtained in this chapter is increased by 10.83 %. Hence, it is decisive for enhancing the productivity of 1,3-PD to optimize the feeding rate of glycerol and the switching instants between the batch and feeding processes in fed-batch fermentation of glycerol to 1,3-PD. The concentration change of 1,3-PD obtained by the optimal strategy is shown in Fig. 7.2. For the purpose of comparison, the experimental data and the 1,3-PD concentration obtained in Chap. 5 are also shown in Fig. 7.2. From Fig. 7.2, we conclude that 1,3-PD concentration at the terminal time in this chapter is actually higher than the ones previously reported.

## 7.6 Conclusion

In this chapter, we investigated the optimal control of a time-dependent switched system arising in fed-batch fermentation. In order to obtain a high concentration of 1,3-PD at the terminal time, an optimal control mode was presented. A computational approach was developed to seek the optimal solution of the constrained optimal control problem. Numerical results showed that the target production concentration actually increased compared with previous results.

# Chapter 8
# Optimal Control of State-Dependent Switched Systems

## 8.1 Introduction

In this chapter, we consider optimal control of state-dependent switched systems in fed-batch process. A proper feeding rate, with the right component constitution, is required in order to improve production during the fed-batch process. This feed should be balanced enough to keep the growth of the microorganism at a desired specific growth rate and reduce simultaneous inhibitory effects by excessive substrate and by-products. To effectively avoid the inhibitory effects and maximize the production of target product at the terminal time in glycerol fed-batch fermentation, we consider the fed-batch process switches between the batch process and the feeding process. Furthermore, if the glycerol concentration drops below the lower switching concentration, then the feeding process is active. While the glycerol concentration rises above upper switching concentration, the feeding process stops and another batch process is active. For a mathematical model, the continuous trajectory evolution observes the batch and the feeding dynamical systems. Note that we must include the distinction between "batch process" and "feeding process." This produces a hybrid state space: involving both discrete and continuous components. This discontinuous alteration of trajectory can be modeled as a state-dependent switched system. Taking the concentration of 1,3-PD at the terminal time as the cost functional, we present an optimal switching control model subject to the switched system and constraints of continuous state inequality and control function. Due to the complex nature of the control problem, it is not possible to derive an analytical solution. Thus, it is necessary to rely on numerical methods for solving the problem. However, because the number of the switchings is not known a priori in the optimal switching control problem, existing methods cannot be used to solve such problem directly. These include the control parameterization technique [109, 240] and the time-scaling transformation [242]. Hence, a new approach is needed for this unconventional optimal control problem.

 
C. Liu, Z. Gong, *Optimal Control of Switched Systems Arising in Fermentation Processes*, Springer Optimization and Its Applications 97,
DOI 10.1007/978-3-662-43793-3_8

In this chapter, we develop a two-level optimization approach. By choosing a number of the switchings, the inner optimization problem becomes a combined optimal parameter selection and optimal control problem which can be handled by the control parameterization technique in conjunction with the time-scaling transformation. The choice of the number of the switching is the outer optimization problem which is solved by a heuristic approach. Numerical results show that by employing the optimal control policy, the concentration of 1,3-PD at the terminal time can be increased considerably and the number of switchings can be cut down greatly.

The main references of this chapter are [149] and [151].

## 8.2 State-Dependent Switched Systems

The whole fed-batch process switches between the batch process and the feeding process. In order to reduce inhibitory effects of excessive substrate for cells growth, glycerol should be kept in a given range. Namely, if the glycerol concentration drops below the lower switching concentration, then the feeding process is active. While the glycerol concentration rises above the upper switching concentration, the feeding process stops and another batch process is active.

Under Assumptions 5.1 and 5.2, mass balances of biomass, substrate, and products in the batch process are written as follows:

$$
\begin{cases}
\dot{x}_1(t) = q_1(\boldsymbol{x}(t))x_1(t),\\
\dot{x}_2(t) = -q_2(\boldsymbol{x}(t))x_1(t),\\
\dot{x}_3(t) = q_3(\boldsymbol{x}(t))x_1(t),\\
\dot{x}_4(t) = q_4(\boldsymbol{x}(t))x_1(t),\\
\dot{x}_5(t) = q_5(\boldsymbol{x}(t))x_1(t),\\
\dot{x}_6(t) = 0,
\end{cases}
\qquad (8.1)
$$
$$\boldsymbol{x}(0) = \boldsymbol{x}_0.$$

But in the course of the feeding process, glycerol and alkali are fed into the fermentor. Mass balances of biomass, substrate, and products in the feeding process are given below:

$$
\begin{cases}
\dot{x}_1(t) = (q_1(\boldsymbol{x}(t)) - D(\boldsymbol{x}(t), u(t)))x_1(t),\\
\dot{x}_2(t) = D(\boldsymbol{x}(t), u(t))\Big(\dfrac{c_{s0}}{1+r} - x_2(t)\Big) - q_2(\boldsymbol{x}(t))x_1(t),\\
\dot{x}_3(t) = q_3(\boldsymbol{x}(t))x_1(t) - D(\boldsymbol{x}(t), u(t))x_3(t),\\
\dot{x}_4(t) = q_4(\boldsymbol{x}(t))x_1(t) - D(\boldsymbol{x}(t), u(t))x_4(t),\\
\dot{x}_5(t) = q_5(\boldsymbol{x}(t))x_1(t) - D(\boldsymbol{x}(t), u(t))x_5(t),\\
\dot{x}_6(t) = (1 + r)u(t).
\end{cases}
\qquad (8.2)
$$

In (8.1) and (8.2), $x_1(t), x_2(t), x_3(t), x_4(t), x_5(t)$, and $x_6(t)$ are the concentrations of biomass, glycerol, 1,3-PD, acetate, and ethanol and the volume of culture fluid at $t$ in fermentor, respectively. Let $\boldsymbol{x}(t) := (x_1(t), x_2(t), x_3(t), x_4(t), x_5(t), x_6(t))^\top \in \mathbb{R}^6_+, t \in [0, T]$, be the state vector, where $T$ is the terminal time of the fermentation. $u(t) \in \mathbb{R}$ is the feeding rate of glycerol in the fed-batch culture. $\boldsymbol{x}_0$ is the initial state. $r$ is the velocity ratio of feeding alkali and glycerol. $c_{s0}$ denotes the initial concentration of glycerol in feed. In addition, $D(\boldsymbol{x}(t), u(t))$ is the dilution rate defined as

$$D(\boldsymbol{x}(t), u(t)) = \frac{(1+r)u(t)}{x_6(t)}. \tag{8.3}$$

The specific growth rate of cells $q_1(\boldsymbol{x}(t))$, the specific consumption rate of substrate $q_2(\boldsymbol{x}(t))$, and the specific formation rates of products $q_\ell(\boldsymbol{x}(t)), \ell = 3, 4, 5$, are expressed as

$$q_1(\boldsymbol{x}(t)) = \frac{\Delta_1 x_2(t)}{x_2(t) + k_1} \prod_{\ell=2}^{5} \left(1 - \frac{x_\ell(t)}{x_\ell^*}\right)^{n_\ell}, \tag{8.4}$$

$$q_2(\boldsymbol{x}(t)) = m_2 + q_1(\boldsymbol{x}(t))Y_2 + \frac{\Delta_2 x_2(t)}{x_2(t) + k_2}, \tag{8.5}$$

$$q_\ell(\boldsymbol{x}(t)) = -m_\ell + q_1(\boldsymbol{x}(t))Y_\ell + \frac{\Delta_\ell x_2(t)}{x_2(t) + k_\ell}, \quad \ell = 3, 4, \tag{8.6}$$

$$q_5(\boldsymbol{x}(t)) = q_2(\boldsymbol{x}(t)) \left(\frac{c_1}{c_2 + q_1(\boldsymbol{x}(t))x_2(t)} + \frac{c_3}{c_4 + q_1(\boldsymbol{x}(t))x_2(t)}\right). \tag{8.7}$$

Under anaerobic conditions at 37 °C and pH 7.0, the critical concentrations for cell growth and the kinetic parameters in (8.4)–(8.7) are as given in Table 5.1.

Let $u(t)$ be the control function. Then, the state-based switched system describing the whole process of fed-batch culture can be formulated as

$$\begin{cases} \dot{\boldsymbol{x}}(t) = \boldsymbol{f}_{j(t)}(\boldsymbol{x}(t), u(t)), \\ u(t) \in U_{j(t)}, \ t \in [0, T], \\ \boldsymbol{x}(0) = \boldsymbol{x}_0, \end{cases} \tag{8.8}$$

where the mapping $j(\cdot) : [0, T] \to \mathcal{J} := \{1, 2\}$ is a piecewise constant function of time, called switching signal. $j(t) = 1$ corresponds to the batch process and $j(t) = 2$ denotes the feeding process. So the forbidden regions may be defined as $\mathcal{R}_1 = \{\boldsymbol{x}(t) \in \mathbb{R}^6_+ | \, x_2(t) < \alpha_*\}$ and $\mathcal{R}_2 = \{\boldsymbol{x}(t) \in \mathbb{R}^6_+ | \, x_2(t) > \alpha^*\}$ with $0 < \alpha_* < \alpha^*$, which respectively denote the lower and the upper switching concentrations of glycerol. Furthermore, switching sets can be defined as $\mathcal{S}_{1,2} = \{\boldsymbol{x}(t) \in \mathbb{R}^6_+ | \, x_2(t) = \alpha_*\}$ and $\mathcal{S}_{2,1} = \{\boldsymbol{x}(t) \in \mathbb{R}^6_+ | \, x_2(t) = \alpha^*\}$. The switching signal satisfies the switching rules below:

($SR_1$) If $\boldsymbol{x}(t) \in \mathcal{R}_k$ for some $t \in [0, T]$ and some $k \in \mathcal{J}$, then $j(t) \neq k$.
($SR_2$) If $\tilde{\tau}$ is a switching instant, then $\boldsymbol{x}(\tilde{\tau}) \in \mathcal{S}_{j(\tilde{\tau}-),j(\tilde{\tau}+)}$.

In addition, the state of the system (8.8) doesn't undergo jump at the switching instants according to the fermentation process.

Let $U_1 := \{0\}$ and $U_2 := [a_*, a^*]$, where $a_*$ and $a^*$ are positive constants which denote the minimal and maximal rates of adding glycerol, respectively. We now define the class of admissible control functions as

$$\mathcal{U} := \left\{u \mid u \in L_\infty([0, T], \mathbb{R}) \text{ and } u(t) \in U_{j(t)} \text{ for } t \in [0, T]\right\}, \tag{8.9}$$

where $L_\infty([0, T], \mathbb{R})$ is the Banach space of all essentially bounded functions from $[0, T]$ into $\mathbb{R}$.

It should be noted that since the concentrations of biomass, glycerol, and products and the volume of culture fluid are restricted in a certain range according to the fermentation process, we consider the properties of the system with state in $\tilde{W} := \prod_{\ell=1}^{6} \left[x_{*\ell}, x_\ell^*\right]$, where $x_{*\ell}$, $x_\ell^*$, $\ell = 1, 2, \ldots, 5$, are as given in Table 5.1, $x_{*,6} = 4$ and $x_6^* = 7$. In addition, the volume of the culture fluid should attain a specific volume $\Delta$ at the terminal time, so the target set $\mathcal{S}_T$ of the solutions can be defined as $\mathcal{S}_T = \{\boldsymbol{x}(T) \mid x_6(T) = \Delta\}$.

For the system (8.8), some important properties are given as follows.

*Property 8.1.* The function $\boldsymbol{f}_j : \mathbb{R}_+^6 \times U_1 \cup U_2 \to \mathbb{R}^6$, $j \in \mathcal{J}$ defined by (8.1) and (8.2) satisfies that

(a) $\boldsymbol{f}_j$ is affine in control $u$,
(b) $\boldsymbol{f}_j$ is continuous on $u$ and $\boldsymbol{x}$,
(c) $\boldsymbol{f}_j$ is of linear growth, that is, there exist two positive constants $\alpha$ and $\beta$ such that the inequality

$$\max\left\{\|\boldsymbol{f}_j(\boldsymbol{x}, u)\| \mid u(t) \in U_j, j = 1, 2\right\} \leq \alpha + \beta\|\boldsymbol{x}\| \tag{8.10}$$

holds.

*Proof.* (a) It is easy to verify that $\boldsymbol{f}_j$ is affine in control $u$ by definition.
(b) We can also conclude that $\boldsymbol{f}_j$ is continuous on $u$ and $\boldsymbol{x}$ by expressions in (8.1) and (8.2).
(c) The proof can be completed using the method similar to the proof of Proposition 5.1 in Chap. 5. □

**Theorem 8.1.** *For any $u \in \mathcal{U}$, the switching system* (8.8) *with initial condition* $(\boldsymbol{x}(0), j(0))$ *consistent with* ($SR_1$) *has a solution* $(\boldsymbol{x}(\cdot|u), j(\cdot|u))$.

*Proof.* The proof is similar to that given for the proof of Theorem 1 in [221]. □

**Theorem 8.2.** *If* $(\boldsymbol{x}(\cdot|u), j(\cdot|u))$ *is a solution of the switching system* (8.8) *with fixed initial condition* $(\boldsymbol{x}(0), j(0))$, *then* $\boldsymbol{x}(\cdot|u)$ *is uniformly bounded and Lipschitzian. Furthermore,* $\boldsymbol{x}(\cdot|u)$ *is compact in* $C(I, \mathbb{R}^6_+)$.

*Proof.* Given the initial condition $(\boldsymbol{x}(0), j(0))$ and for all $u \in \mathcal{U}$, the solution of the system (8.8) can be written as

$$\boldsymbol{x}(t|u) = \boldsymbol{x}(0) + \int_0^t \boldsymbol{f}_{j(s|u)}(\boldsymbol{x}(s|u), u(s))\mathrm{d}s, \ \forall t \in [0, T].$$

From (8.10), we have

$$\|\boldsymbol{x}(t|u)\| \leqslant \|\boldsymbol{x}(0)\| + \int_0^t (\alpha + \beta\|\boldsymbol{x}(s|u)\|)\mathrm{d}s.$$

By applying Lemma 4.1, we must obtain

$$\|\boldsymbol{x}(t|u)\| \leqslant C\exp(\beta T), \forall t \in [0, T], \tag{8.11}$$

with $C := \|\boldsymbol{x}(0)\| + \alpha T$. Moreover, for all $t, t' \in [0, T]$,

$$\|\boldsymbol{x}(t|u) - \boldsymbol{x}(t'|u)\| \leqslant \left| \int_{t'}^t (\alpha + \beta\|\boldsymbol{x}(s|u)\|)\mathrm{d}s \right|.$$

Letting $L := \alpha + \beta C\exp(\beta T)$, we conclude that

$$\|\boldsymbol{x}(t|u) - \boldsymbol{x}(t'|u)\| \leqslant L|t - t'|. \tag{8.12}$$

In view of (8.1) and (8.2), we may rewrite (8.8) as differential inclusion: $\boldsymbol{F}_j(\boldsymbol{x}(t)) := \{\boldsymbol{f}_j(\boldsymbol{x}, u)\}_{u(t)\in U_j}$. Since $U_j$ is convex and compact in each interswitching interval, we must conclude that $\boldsymbol{x}(\cdot)$ is compact in $C(I, \mathbb{R}^6_+)$. □

**Theorem 8.3.** *If* $(\boldsymbol{x}(\cdot|u), j(\cdot|u))$ *is a solution of the system* (8.8) *with fixed initial condition, then* $j(\cdot|u)$ *is piecewise constant with finitely many switchings.*

*Proof.* In interswitching intervals $(\tau_k, \tau_{k+1})$, the relevant switching sets $\mathcal{S}_{1,2}, \mathcal{S}_{2,1}$ are compact and disjoint. So the trajectories at switching instants $\tau_k$ and $\tau_{k+1}$, i.e., $\boldsymbol{x}(\tau_k)$ and $\boldsymbol{x}(\tau_{k+1})$, satisfy $\|\boldsymbol{x}(\tau_{k+1}) - \boldsymbol{x}(\tau_k)\| > 0$. By Theorem 8.2, we conclude that there exists a positive constant $L$ such that

$$\|\boldsymbol{x}(\tau_{k+1}) - \boldsymbol{x}(\tau_k)\| \leqslant L|\tau_{k+1} - \tau_k|.$$

Then, a positive lower bound on the length of interval $(\tau_k, \tau_{k+1})$ can be obtained. As a result, we must obtain that the switching times are finite. □

*Remark 8.1.* Theorem 8.3 ensures the switching system (8.8) has only a finite number of switchings. In fact, all software packages currently available to model such a class of dynamical systems are developed for situations where no infinite number of switchings exists. For if this is not the case, then it will cause a major difficulty in numerical computation. Furthermore, it is also highly undesirable in the actual fed-batch fermentation process.

## 8.3 Optimal Control Models

For mathematical convenience, define the set of the system (8.8), $\mathcal{S}_0$, as follows:

$$\mathcal{S}_0 = \{(\boldsymbol{x}(\cdot|u), j(\cdot|u))|\ (\boldsymbol{x}(t|u), j(t|u)) \text{ is the solution to the system (8.8) with } u \in \mathcal{U} \text{ for all } t \in [0, T]\}.$$

Since the concentrations of biomass, glycerol, and products are restricted in $W$ and the volume of the culture fluid should attain a specific volume $\Delta$ at the terminal time, we denote the admissible set of the solutions by

$$\mathcal{S} = \left\{(\boldsymbol{x}(\cdot|u), j(\cdot|u)) \in \mathcal{S}_0|\ \boldsymbol{x}(t|u) \in \tilde{W} \text{ for all } t \in [0, T] \text{ and } \boldsymbol{x}(T|u) \in \mathcal{S}_T\right\}.$$

Furthermore, the set of the feasible control functions can be defined as

$$\mathcal{F} = \{u \in \mathcal{U}|\ (\boldsymbol{x}(\cdot|u), j(\cdot|u)) \in \mathcal{S}\}.$$

With the knowledge of the above definitions, the problem of optimizing the feeding rate of glycerol such that the concentration of 1,3-PD at the terminal time as high as possible can be described as follows:

$$\text{(SOCP)} \qquad \min\ J(u) = -x_3(T|u)$$
$$\text{s.t.}\ \ u \in \mathcal{F}.$$

To discuss the existence of optimal control for (SOCP), we consider the weak convergence in $L_\infty([0, T], \mathbb{R})$ for control function $u(\cdot)$ and uniform convergence in $C([0, T], \mathbb{R}^6)$ for the trajectory $\boldsymbol{x}(\cdot)$, and $j(\cdot)$ converges in $[0, T]$ to switching instant $\tilde{\tau}$.

**Theorem 8.4.** *The set* $\mathcal{S}$ *of admissible solutions defined in* (8.13) *is compact.*

*Proof.* Let $\left\{\tilde{s}^i := (\boldsymbol{x}^i(\cdot), j^i(\cdot))\right\} \subseteq \mathcal{S}$ be an admissible solution sequence of (8.8) and $u^i(\cdot)$ be the corresponding control such that $\boldsymbol{x}^i(\cdot) = \boldsymbol{x}^i\left(\cdot|u^i\right)$ and $j^i(\cdot) = j^i\left(\cdot|u^i\right)$. The boundedness of control function permits extraction of a subsequence $\left\{u^{i_k}\right\} \subseteq \left\{u^i\right\}$ for which $u^{i_k}$ weakly converges to $u$. We can also extract $\left\{\tau_v^{i_k}\right\} \subseteq \left\{\tau_v^i\right\}$ that converges $\tau_v$ in $[0, T]$ for each $v$. Thus, $j^{i_k} \to j$. Moreover, in view of Theorem 8.2, we can also extract $\boldsymbol{x}^{i_k} \to \boldsymbol{x}$ uniformly on $[0, T]$. Let $\tilde{s} := (\boldsymbol{x}(\cdot), j(\cdot)) \in \mathcal{S}$. Then, we obtain that $\boldsymbol{x}(t) \in \tilde{W}$ and $\boldsymbol{x}(T) \in \mathcal{S}_T$ since $\tilde{W}$ and the target set $\mathcal{S}_T$ are closed. Furthermore, $\{\tilde{s}^i\}$ is regular, so we consider an open interval $\mathcal{I} \subseteq [\tau_v, \tau_{v+1}]$, which implies $\mathcal{I} \subseteq [\tau_v^i, \tau_{v+1}^i]$ and $j^i(\cdot) \equiv k$ ($k$ is fixed) on $\mathcal{I}$. Applying the convergence theorem in [6] and Property 8.1, we conclude that $\tilde{s}$ satisfies (8.8). Finally, consider some $t$ in $\mathcal{I}$ as above for which $j(\cdot) \equiv k$. Since $\mathcal{R}_k$ is open, $\lim_{k\to\infty} \boldsymbol{x}^{i_k}(t) = \boldsymbol{x}(t) \notin \mathcal{R}_k$, which implies (SR$_1$). For a switching time $t = \tau_v$, we have $\tau_v^{i_k} \to \tau_v$ and $\boldsymbol{x}^{i_k}(\tau_v^{i_k}) \to \boldsymbol{x}(\tau_v)$. Since the switching sets $\mathcal{S}_{j(\tau_v -), j(\tau_v +)}$ is closed, this gives $\boldsymbol{x}(\tau_v)$ in the appropriate switching set, which verifies (SR$_2$). Thus, $\tilde{s} \in \mathcal{S}_0$ and so $\tilde{s} \in \mathcal{S}$. □

**Theorem 8.5.** (SOCP) *has at least one optimal solution.*

*Proof.* Since $J$ is continuous in $\boldsymbol{x}$ and the admissible solutions set $\mathcal{S}$ is compact, (SOCP) has at least one optimal solution. □

Although one can directly control the system (8.8) by the control function, a switching sequence will be generated implicitly along with the evolution of the system state trajectory. This makes that existing methods cannot be used to solve such a problem directly. Hence, we reformulate (SOCP) as a combined optimal parameter selection and optimal control problem by introducing additional state constraints at the switchings.

Let $\tau_i, i = 0, 1, \ldots, N+1, N \in \mathbb{Z}_+$, be the switching instants such that

$$0 = \tau_0, \tau_{i-1} < \tau_i, i = 1, 2, \ldots, N+1, \text{ and } \tau_{N+1} = T, \tag{8.13}$$

where $N$ is the number of the switchings and $\mathbb{Z}_+$ is the set of positive integers. The corresponding vector of the switching instants is

$$\boldsymbol{\tau} := (\tau_1, \tau_2, \ldots, \tau_N)^\top.$$

Denote respectively the admissible regions of systems (8.1) and (8.2) as

$$g_1(\boldsymbol{x}(t|u)) = x_2(t|u) - \alpha_* \geqslant 0 \text{ and } g_2(\boldsymbol{x}(t|u)) = \alpha^* - x_2(t|u) \geqslant 0.$$

Let

$$h_\ell(\boldsymbol{x}(t|u)) = x_\ell(t|u) - x_{*\ell},$$

$$h_{6+\ell}(\boldsymbol{x}(t|u)) = x_\ell^* - x_\ell(t|u), \ \ell = 1, 2, \ldots, 6.$$

Then, the condition $\boldsymbol{x}(t|u) \in \tilde{W}$ is transformed into

$$h_l(\boldsymbol{x}(t|u)) \geqslant 0, \ l = 1, 2, \ldots, 12.$$

Furthermore, let $\phi_i(u) = (x_2(\tau_i|u) - \alpha_*)(x_2(\tau_i|u) - \alpha^*)$.

The switching rule ($SR_2$) is equivalent to

$$\phi_i(u) = 0. \tag{8.14}$$

Furthermore, let $\phi_{N+1}(u) = x_6(T|u) - \Delta$. Then, the constraint $\boldsymbol{x}(T|u) \in \mathcal{S}_T$ is equivalently transcribed into

$$\phi_{N+1}(u) = 0. \tag{8.15}$$

Now, (SOCP) can be reformulated as a combined parameter selection and optimal control problem:

$$\begin{aligned}
\text{(PSOCP)} \qquad & \min\ J(u, \boldsymbol{\tau}, N) = -x_3(T|u) \\
& \text{s.t. } g_{j(t|u)}(\boldsymbol{x}(t|u)) \geqslant 0, \\
& \quad h_l(\boldsymbol{x}(t|u)) \geqslant 0,\ l = 1, 2, \ldots, 12, \\
& \quad \phi_i(u) = 0, \\
& \quad u(t) \in U_{j(t)}, \\
& \quad t \in (\tau_{i-1}, \tau_i],\ i = 1, 2, \ldots, N+1, \\
& \quad N \in \mathbb{Z}_+.
\end{aligned}$$

Let $\Gamma$ be the set of all vectors $\boldsymbol{\tau}$ such that (8.13) and (8.14) are satisfied. Then, combined optimal parameter selection and optimal control problem (PSOCP) can be viewed as a two-level optimization problem as follows:

$$\min_{N \in \mathbb{Z}_+} \min_{u \times \boldsymbol{\tau} \in \mathcal{F} \times \Gamma} J(u, \boldsymbol{\tau}, N).$$

To be more specific, define the inner optimization problem as

$$(\text{IOP}_N) \qquad \hat{J}(N) = \min_{u \times \boldsymbol{\tau} \in \mathcal{F} \times \Gamma} J(u, \boldsymbol{\tau}, N)$$

for each $N \in \mathbb{Z}_+$.

For a given $N \in \mathbb{Z}_+$, there exists an optimal solution of the inner optimization problem ($\text{IOP}_N$) by Theorem 8.5, which is denoted by $(u^*(\cdot|N), \boldsymbol{\tau}^*(N))$. As a result, (PSOCP) is equivalent to

$$\min_{N \in \mathbb{Z}_+} J(u^*(\cdot|N), \boldsymbol{\tau}^*(N)).$$

To solve the problem (PSOCP), we should firstly solve the inner optimization problem ($\text{IOP}_N$) and then optimize the number of switching $N$. In the numerical computation, we use a heuristic approach to determine this number. We start by fixing the number of switchings to be a fixed integer $N$ and solve ($\text{IOP}_N$). Then, we increase the number of switchings from $N$ to $N+d$ (where $d$ is an integer) and solve the corresponding (IOP$_N$) again. If there is no decrease in the optimal cost functional, we take the previous value $N$ to be the number of switchings. Consequently, a solution method for (IOP$_N$) is necessary to be developed.

## 8.4 Solution Methods for the Inner Optimization Problem

In this section, we shall use the control parameterization method [109, 240] and time-scaling transformation [242] to develop a numerical method to solve (IOP$_N$).

For each $p_i \geqslant 1, i \in \{1,2,\ldots,N+1\}$, let the time subinterval $[\tau_{i-1}, \tau_i]$ be partitioned into $n_{p_i}$ subintervals with $n_{p_i}+1$ partition points denoted by

$$\tau_0^{p_i}, \tau_1^{p_i}, \ldots, \tau_{n_{p_i}}^{p_i}, \tau_0^{p_i} = \tau_{i-1}, \tau_{n_{p_i}}^{p_i} = \tau_i, \text{ and } \tau_{k-1}^{p_i} \leqslant \tau_k^{p_i}.$$

Let $n_{p_i}$ be chosen such that $n_{p_i+1} \geqslant n_{p_i}$. The control is now approximated in the form of a piecewise constant function as follows:

$$u^p(t) = \sum_{i=1}^{N+1} \sum_{k=1}^{n_{p_i}} \sigma^{p_i,k} \chi_{(\tau_{k-1}^{p_i}, \tau_k^{p_i}]}(t). \tag{8.16}$$

Here, $\chi_{(\tau_{k-1}^{p_i}, \tau_k^{p_i}]}$ is the indicator function on the interval $\left(\tau_{k-1}^{p_i}, \tau_k^{p_i}\right]$ defined by

$$\chi_{(\tau_{k-1}^{p_i}, \tau_k^{p_i}]}(t) = \begin{cases} 1, & t \in \left(\tau_{k-1}^{p_i}, \tau_k^{p_i}\right], \\ 0, & \text{otherwise.} \end{cases}$$

Let $\kappa := \sum_{i=1}^{N+1} n_{p_i}$. Then, $\boldsymbol{\sigma}^p := ((\boldsymbol{\sigma}^{p_1})^\top, \ldots, (\boldsymbol{\sigma}^{p_{N+1}})^\top)^\top \in \mathbb{R}^\kappa$, where $\boldsymbol{\sigma}^{p_i} := (\sigma^{p_i,1}, \ldots, \sigma^{p_i,n_{p_i}})^\top$ defines the heights of the approximate control (8.16). From (8.9), it is clear that

$$\sigma^{p_i,k} = 0, \text{ if } u(t) \in U_1, \text{ and } a_* \leqslant \sigma^{p_i,k} \leqslant a^*, \text{ if } u(t) \in U_2, \tag{8.17}$$

for $k = 1, \ldots, n_{p_i}; i = 1, \ldots, N+1$. Let $\Xi^p$ be the set of all those $\boldsymbol{\sigma}^p$ which satisfy the constraints (8.17). Here, $\sigma^{p_i,k}$ and the time points $\tau_k^{p_i}$, $k = 1,2,\ldots,n_{p_i}; i = 1,2,\ldots,N+1$, are decision variables.

However, (IOP$_N$) with time points $\tau_k^{p_i}$, $k = 1,2,\ldots,n_{p_i}; i = 1,2,\ldots,N+1$, taken as the decision variables will encounter numerical difficulties as mentioned in Chap. 7. For this reason, we introduce a time-scaling transformation as in Chap. 7 to map these variable time points into preassigned fixed knots in a new time scale. It is achieved by introducing a transform from $t \in [0,T]$ to $s \in [0, N+1]$ as follows:

$$\dot{t}(s) = v^p(s), \quad t(0) = 0, \tag{8.18}$$

where $v^p$ is given by

$$v^p(s) = \sum_{i=1}^{N+1} \sum_{k=1}^{n_{p_i}} \delta_k^{p_i} \chi_{\left(i-1+\frac{k-1}{n_{p_i}},\, i-1+\frac{k}{n_{p_i}}\right]}(s). \tag{8.19}$$

In (8.19), $\delta_k^{p_i} \geqslant 0, k = 1, 2, \ldots, n_{p_i}; i = 1, 2, \ldots, N+1$, are decision variables. Let $\boldsymbol{\delta}^p$ be the vector whose components are $\delta_k^{p_i}, k = 1, 2, \ldots, n_{p_i}; i = 1, 2, \ldots, N+1$, and $\Omega^p$ be the set of all such $\boldsymbol{\delta}^p$. Furthermore, denote the set of all $v^p$ obtained by elements from $\Omega^p$ via (8.19) as $\mathcal{V}^p$. Clearly, each $v^p \in \mathcal{V}^p$ is determined uniquely by a $\boldsymbol{\delta}^p \in \Omega^p$ and vice versa. Thus, we write $v^p(\cdot)$ as $v^p(\cdot|\boldsymbol{\delta}^p)$.

Let

$$w^p(s) = u^p(t(s)).$$

Then

$$w^p(s) = \sum_{i=1}^{N+1} \sum_{k=1}^{n_{p_i}} \sigma^{p_i,k} \chi_{\left(i-1+\frac{k-1}{n_{p_i}},\, i-1+\frac{k}{n_{p_i}}\right]}(s).$$

Since $w^p$ is determined uniquely by $\boldsymbol{\sigma}^p$ and vice versa, it is written as $w^p(\cdot|\boldsymbol{\sigma}^p)$.

Define

$$\tilde{\boldsymbol{x}}(s) := (\boldsymbol{x}(s)^\top, t(s))^\top, \quad \tilde{j}(s) := j(t(s))$$

and

$$\tilde{\boldsymbol{f}}_{\tilde{j}(s)}(\tilde{\boldsymbol{x}}(s), \boldsymbol{\sigma}^p, \boldsymbol{\delta}^p) := ((v^p(s)\boldsymbol{f}_{j(t(s))}(\boldsymbol{x}(t(s)), w(s|\boldsymbol{\sigma}^p))^\top, v^p(s|\boldsymbol{\delta}^p))^\top.$$

Let $\tilde{\boldsymbol{x}}(\cdot|\boldsymbol{\sigma}^p, \boldsymbol{\delta}^p)$ be the solution of the following system corresponding to the control parameter vector $(\boldsymbol{\sigma}^p, \boldsymbol{\delta}^p) \in \Xi^p \times \Omega^p$:

$$\begin{cases} \dot{\tilde{\boldsymbol{x}}}(s) = \tilde{\boldsymbol{f}}_{\tilde{j}(s)}(\tilde{\boldsymbol{x}}(s), \boldsymbol{\sigma}^p, \boldsymbol{\delta}^p), \\ \tilde{\boldsymbol{x}}(0) = (\boldsymbol{x}_0^\top, 0)^\top. \end{cases}$$

Then, for each $p$, ($\text{IOP}_N$) is transformed into a standard parameter selection problem as follows:

$$(\text{IOP}_N(\text{p})) \qquad \min J(\boldsymbol{\sigma}^p, \boldsymbol{\delta}^p) = -\tilde{x}_3(N+1|\boldsymbol{\sigma}^p, \boldsymbol{\delta}^p)$$

$$\text{s.t.} \quad g_{\tilde{j}(s|\boldsymbol{\sigma}^p,\boldsymbol{\delta}^p)}(\tilde{\boldsymbol{x}}(s|\boldsymbol{\sigma}^p, \boldsymbol{\delta}^p)) \geqslant 0, \tag{8.20}$$

$$h_l(\tilde{\boldsymbol{x}}(s|\boldsymbol{\sigma}^p, \boldsymbol{\delta}^p)) \geqslant 0, \; l = 1, 2, \ldots, 12, \tag{8.21}$$

$$\phi_i(\boldsymbol{\sigma}^p, \boldsymbol{\delta}^p) = 0, \tag{8.22}$$
$$\boldsymbol{\sigma}^p \in \Xi^p,$$
$$\boldsymbol{\delta}^p \in \Omega^p,$$
$$s \in (i-1, i],\ i = 1, 2, \ldots, N+1.$$

Since constraints (8.20) and (8.21) in ($\text{IOP}_N$(p)) are continuous inequality constraints, we shall use the method in Chap. 6 to deal with these continuous inequality constraints. Let

$$G(\boldsymbol{\sigma}^p, \boldsymbol{\delta}^p) := \sum_{l=1}^{12} \int_0^{N+1} \min\{0, h_l(\tilde{\boldsymbol{x}}(s|\boldsymbol{\sigma}^p, \boldsymbol{\delta}^p))\}\,\mathrm{d}s + \int_0^{N+1} \min\left\{0, g_{\tilde{j}(s|\boldsymbol{\sigma}^p,\boldsymbol{\delta}^p)}(\tilde{\boldsymbol{x}}(s|\boldsymbol{\sigma}^p, \boldsymbol{\delta}^p))\right\}\mathrm{d}s.$$

Then, conditions $g_{\tilde{j}(s)}(\tilde{\boldsymbol{x}}(s|\boldsymbol{\sigma}^p, \boldsymbol{\delta}^p)) \geqslant 0$ and $h_l(\tilde{\boldsymbol{x}}(s|\boldsymbol{\sigma}^p, \boldsymbol{\delta}^p)) \geqslant 0$, $l = 1, 2, \ldots, 12$, are equivalently transcribed into

$$G(\boldsymbol{\sigma}^p, \boldsymbol{\delta}^p) = 0. \tag{8.23}$$

However, $G(\cdot,\cdot)$ is non-differentiable at the points $h_l = 0$ and $g_{\tilde{j}} = 0, l \in \{1, 2, \ldots, 12\}; \tilde{j} \in \{1, 2\}$. We replace (8.23) with

$$\tilde{G}_{\varepsilon,\gamma}(\boldsymbol{\sigma}^p, \boldsymbol{\delta}^p) = \gamma + \sum_{l=1}^{12} \int_0^{N+1} \varphi_\varepsilon(h_l(\tilde{\boldsymbol{x}}(s|\boldsymbol{\sigma}^p, \boldsymbol{\delta}^p)))\mathrm{d}s + \int_0^{N+1} \varphi_\varepsilon\left(g_{\tilde{j}(s|\boldsymbol{\sigma}^p,\boldsymbol{\delta}^p)}(\tilde{\boldsymbol{x}}(s|\boldsymbol{\sigma}^p, \boldsymbol{\delta}^p))\right)\mathrm{d}s \geqslant 0, \tag{8.24}$$

where $\varepsilon > 0$, $\gamma > 0$ and

$$\varphi_\varepsilon(\eta) = \begin{cases} \eta, & \text{if } \eta < -\varepsilon, \\ -\dfrac{(\eta-\varepsilon)^2}{4\varepsilon}, & \text{if } -\varepsilon \leqslant \eta \leqslant \varepsilon, \\ 0, & \text{if } \eta > \varepsilon. \end{cases} \tag{8.25}$$

Thus, ($\text{IOP}_N$(p)) is approximated by a sequence of nonlinear programming problems ($\text{IOP}_{N,\varepsilon,\gamma}$(p)) defined by replacing constraint (8.23) with (8.24). We shall use the improved particle swarm optimization (PSO) algorithm developed in Chap. 5 to solve ($\text{IOP}_{N,\varepsilon,\gamma}$(p)).

To solve ($\text{IOP}_{N,\varepsilon,\gamma}$(p)) by the improved PSO algorithm, the gradient formulae given in the following theorems are needed.

**Theorem 8.6.** *For the constraint $\tilde{G}_{\varepsilon,\gamma}(\boldsymbol{\sigma}^p, \boldsymbol{\delta}^p)$ given in (8.24), it holds that its gradients with respect to parameterized control $\boldsymbol{\sigma}^p$ and $\boldsymbol{\delta}^p$ are*

$$\frac{\partial \tilde{G}_{\varepsilon,\gamma}(\boldsymbol{\sigma}^p, \boldsymbol{\delta}^p)}{\partial \sigma^{p_i,k}} = \int_{i-1+\frac{k-1}{n_{p_i}}}^{i-1+\frac{k}{n_{p_i}}} \frac{\partial \tilde{H}(\tilde{\boldsymbol{x}}(s|\boldsymbol{\sigma}^p, \boldsymbol{\delta}^p), \boldsymbol{\sigma}^p, \boldsymbol{\delta}^p, \tilde{\boldsymbol{\lambda}}(s))}{\partial \sigma^{p_i,k}} \mathrm{d}s,$$

*and*

$$\frac{\partial \tilde{G}_{\varepsilon,\gamma}(\boldsymbol{\sigma}^p, \boldsymbol{\delta}^p)}{\partial \delta^{p_i,k}} = \int_{i-1+\frac{k-1}{n_{p_i}}}^{i-1+\frac{k}{n_{p_i}}} \frac{\partial \tilde{H}(\tilde{\boldsymbol{x}}(s|\boldsymbol{\sigma}^p, \boldsymbol{\delta}^p), \boldsymbol{\sigma}^p, \boldsymbol{\delta}^p, \tilde{\boldsymbol{\lambda}}(s))}{\partial \delta^{p_i,k}} \mathrm{d}s,$$

*where*

$$\begin{aligned}\tilde{H}\Big(\tilde{\boldsymbol{x}}\,(s|\boldsymbol{\sigma}^p, \boldsymbol{\delta}^p), \boldsymbol{\sigma}^p, \boldsymbol{\delta}^p, \tilde{\boldsymbol{\lambda}}(s)\Big) &= \sum_{l=1}^{12} \varphi_\varepsilon(h_l(\tilde{\boldsymbol{x}}(s|\boldsymbol{\sigma}^p, \boldsymbol{\delta}^p))) + \varphi_\varepsilon\Big(g_{\tilde{j}(s)}(\tilde{\boldsymbol{x}}(s|\boldsymbol{\sigma}^p, \boldsymbol{\delta}^p))\Big) \\ &\quad + \tilde{\boldsymbol{\lambda}}^\top(s)\tilde{\boldsymbol{f}}_{\tilde{j}(s|\boldsymbol{\sigma}^p,\boldsymbol{\delta}^p)}(\tilde{\boldsymbol{x}}(s|\boldsymbol{\sigma}^p, \boldsymbol{\delta}^p), \boldsymbol{\sigma}^p, \boldsymbol{\delta}^p),\end{aligned}$$

*and*

$$\tilde{\boldsymbol{\lambda}}(s) = (\tilde{\lambda}_1(s), \tilde{\lambda}_2(s), \tilde{\lambda}_3(s), \tilde{\lambda}_4(s), \tilde{\lambda}_5(s), \tilde{\lambda}_6(s), \tilde{\lambda}_7(s))^\top$$

*is the solution of the costate system*

$$\dot{\tilde{\boldsymbol{\lambda}}}(s) = -\left(\frac{\partial \tilde{H}(\tilde{\boldsymbol{x}}(s|\boldsymbol{\sigma}^p, \boldsymbol{\delta}^p), \boldsymbol{\sigma}^p, \boldsymbol{\delta}^p, \tilde{\boldsymbol{\lambda}}(s))}{\partial \tilde{\boldsymbol{x}}}\right)^\top,$$

*with the boundary conditions*

$$\begin{aligned}\tilde{\boldsymbol{\lambda}}(N+1) &= (0,0,0,0,0,0,0)^\top, \\ \tilde{\boldsymbol{\lambda}}(0) &= (0,0,0,0,0,0,0)^\top, \\ \tilde{\boldsymbol{\lambda}}(\iota+) &= \tilde{\boldsymbol{\lambda}}(\iota-), \quad \iota = 1, 2, \ldots, N.\end{aligned}$$

*Proof.* The proof can be completed using the method of Chapter 3 in [38]. □

**Theorem 8.7.** *For the constraints $\phi_\iota(\boldsymbol{\sigma}^p, \boldsymbol{\delta}^p)$, $\iota = 1, 2, \ldots, N$, given in (8.22), it holds that their gradients with respect to parameterized control $\boldsymbol{\sigma}^p$ and $\boldsymbol{\delta}^p$ are, respectively,*

$$\frac{\partial \phi_\iota(\boldsymbol{\sigma}^p, \boldsymbol{\delta}^p)}{\partial \sigma^{p_i,k}} = \int_{i-1+\frac{k-1}{n_{p_i}}}^{i-1+\frac{k}{n_{p_i}}} \frac{\partial \bar{H}_\iota(\tilde{\boldsymbol{x}}(s|\boldsymbol{\sigma}^p, \boldsymbol{\delta}^p), \boldsymbol{\sigma}^p, \boldsymbol{\delta}^p, \bar{\boldsymbol{\lambda}}_\iota(s))}{\partial \sigma^{p_i,k}} \mathrm{d}s,$$

*and*

$$\frac{\partial \phi_\iota(\boldsymbol{\sigma}^p, \boldsymbol{\delta}^p)}{\partial \delta^{p_i,k}} = \int_{i-1+\frac{k-1}{n_{p_i}}}^{i-1+\frac{k}{n_{p_i}}} \frac{\partial \bar{H}_\iota(\tilde{\boldsymbol{x}}(s|\boldsymbol{\sigma}^p, \boldsymbol{\delta}^p), \boldsymbol{\sigma}^p, \boldsymbol{\delta}^p, \bar{\boldsymbol{\lambda}}_\iota(s))}{\partial \delta^{p_i,k}} \mathrm{d}s,$$

*where*

$$\bar{H}_\iota(\tilde{\boldsymbol{x}}(s|\boldsymbol{\sigma}^p, \boldsymbol{\delta}^p), \boldsymbol{\sigma}^p, \boldsymbol{\delta}^p, \bar{\boldsymbol{\lambda}}_\iota(s)) = \bar{\boldsymbol{\lambda}}_\iota^\top(s) \tilde{\boldsymbol{f}}_{\tilde{j}(s|\boldsymbol{\sigma}^p, \boldsymbol{\delta}^p)}(\tilde{\boldsymbol{x}}(s|\boldsymbol{\sigma}^p, \boldsymbol{\delta}^p), \boldsymbol{\sigma}^p, \boldsymbol{\delta}^p),$$

*and*

$$\bar{\boldsymbol{\lambda}}_\iota(s) = (\bar{\lambda}_{\iota,1}(s), \bar{\lambda}_{\iota,2}(s), \bar{\lambda}_{\iota,3}(s), \bar{\lambda}_{\iota,4}(s), \bar{\lambda}_{\iota,5}(s), \bar{\lambda}_{\iota,6}(s), \bar{\lambda}_{\iota,7}(s))^\top$$

*is the solution of the costate system*

$$\dot{\bar{\boldsymbol{\lambda}}}_\iota(s) = -\left(\frac{\partial \bar{H}_\iota(\tilde{\boldsymbol{x}}(s|\boldsymbol{\sigma}^p, \boldsymbol{\delta}^p), \boldsymbol{\sigma}^p, \boldsymbol{\delta}^p, \bar{\boldsymbol{\lambda}}_\iota(s))}{\partial \tilde{\boldsymbol{x}}}\right)^\top,$$

*with the boundary conditions*

$$\begin{aligned} \bar{\boldsymbol{\lambda}}_\iota(\iota) &= (0, 2\tilde{x}_2(\iota|\boldsymbol{\sigma}^p, \boldsymbol{\delta}^p) - (\alpha_* + \alpha^*), 0, 0, 0, 0, 0)^\top, \\ \bar{\boldsymbol{\lambda}}_\iota(0) &= (0, 0, 0, 0, 0, 0, 0)^\top, \\ \bar{\boldsymbol{\lambda}}_\iota(\varsigma+) &= \bar{\boldsymbol{\lambda}}_\iota(\varsigma-), \quad \varsigma = 1, 2, \ldots, \iota - 1. \end{aligned}$$

*Proof.* The proof can be completed using the method of Chapter 3 in [38]. □

On the basis of Theorems 8.6 and 8.7, we can obtain an approximately optimal control and switching instants for ($\text{IOP}_N$) as shown in the following algorithm.

**Algorithm 8.1.**

**Step 1.** Choose initial values of $\varepsilon$ and $\gamma$; set parameters $0 < \alpha < 1$ and $0 < \beta < 1$.
**Step 2.** Solve ($\text{IOP}_{N,\varepsilon,\gamma}$(p)) using the improved PSO algorithm to give $(\boldsymbol{\sigma}_{\varepsilon,\gamma}^{p,*}, \boldsymbol{\delta}_{\varepsilon,\gamma}^{p,*})$.
**Step 3.** Check feasibility of $g_{\tilde{j}(s)}(\tilde{\boldsymbol{x}}(s|\boldsymbol{\sigma}_{\varepsilon,\gamma}^{p,*}, \boldsymbol{\delta}_{\varepsilon,\gamma}^{p,*})) \geqslant 0$ and $h_l(\tilde{\boldsymbol{x}}(s|\boldsymbol{\sigma}_{\varepsilon,\gamma}^{p,*}, \boldsymbol{\delta}_{\varepsilon,\gamma}^{p,*})) \geqslant 0$ for $s \in [0, N+1], l = 1, 2, \ldots, 12,$.

**Step 4.** If $(\sigma_{\varepsilon,\gamma}^{p,*}, \delta_{\varepsilon,\gamma}^{p,*})$ is feasible, then go to Step 5. Otherwise, set $\gamma := \alpha\gamma$. If $\gamma < \bar{\gamma}$, where $\bar{\gamma}$ is a prespecified positive constant, then go to Step 6. Otherwise, go to Step 2.

**Step 5.** Set $\varepsilon := \beta\varepsilon$. If $\varepsilon > \bar{\varepsilon}$, where $\bar{\varepsilon}$ is a prespecified positive constant, then go to Step 3. Otherwise, go to Step 6.

**Step 6.** If $\min_{i\in\{1,2,\ldots,N+1\}} n_{p_i} \geq P$, where $P$ is a predefined positive constant, then go to Step 7. Otherwise, go to Step 1 with $n_{p_i}$ increased to $n_{p_i+1}$ for each $i$.

**Step 7.** Construct $u^{p,*}$ and $\boldsymbol{\tau}^{p,*}$ from $(\sigma_{\varepsilon,\gamma}^{p,*}, \delta_{\varepsilon,\gamma}^{p,*})$ and stop.

Then, $(u^{p,*}, \boldsymbol{\tau}^{p,*})$ obtained is an approximately optimal solution of (IOP$_N$).

## 8.5 Numerical Results

In the numerical simulation, the parameters needed in the computation of the solution to (8.8) are listed in the Table 8.1. In Algorithm 8.1, the number of initial particles swarm $N^p$; the maximal iteration $M^p$; the parameters $c_1^p$, $c_2^p$, $P_{cr}$, $M_1^p$, $M_2^p$, $\varepsilon_p$, which have the same meanings as those given in Chap. 5; parameters $\varepsilon$, $\gamma$, $P$, $\alpha$, $\beta$, $\bar{\varepsilon}$, and $\bar{\gamma}$; the initial number of the switchings $N$; and $d$ used for numerical simulation are presented in the Table 8.2. These parameters are derived empirically after numerous experiments.

**Table 8.1** Parameters in the computation of the solution to the system (8.8)

| Parameters | Values |
|---|---|
| $\boldsymbol{x}_0$ | $(0.1115\,\mathrm{g\,L^{-1}}, 495\,\mathrm{mmol\,L^{-1}}, 0, 0, 0, 5\,\mathrm{L})^\top$ |
| $r$ | 0.75 |
| $c_{s0}$ | 10,762 mmol $\mathrm{L^{-1}}$ |
| $a^*$ | $3.058 \times 10^{-4}\,\mathrm{L\,s^{-1}}$ |
| $a_*$ | $1.594 \times 10^{-4}\,\mathrm{L\,s^{-1}}$ |
| $\alpha^*$ | 326.0870 mmol $\mathrm{L^{-1}}$ |
| $\alpha_*$ | 217.3913 mmol $\mathrm{L^{-1}}$ |
| $\Delta$ | 6.55 L |
| $T$ | 24.16 h |

**Table 8.2** Parameters used for numerical simulation

| $N^p$ | $M^p$ | $c_1^p$ | $c_2^p$ | $P_{cr}$ | $M_1^p$ | $M_2^p$ | $\varepsilon_p$ | $P$ | $\varepsilon$ | $\gamma$ | $\alpha$ | $\beta$ | $\bar{\varepsilon}$ | $\bar{\gamma}$ | $N$ | $d$ |
|---|---|---|---|---|---|---|---|---|---|---|---|---|---|---|---|---|
| 150 | 100 | 2 | 2 | 0.5 | 50 | 20 | $10^{-3}$ | 5 | 0.1 | 0.1 | 0.1 | 0.1 | $10^{-5}$ | $10^{-8}$ | 20 | 1 |

The ODEs in the computation process are numerically calculated by improved Euler method with the relative error tolerance $10^{-4}$. All computations are performed in Visual C++ 6.0, and numerical results are plotted by Matlab 7.10.0 on an AMD Athlon 64 X2 Dual Core Processor TK-57 1.90 GHz machine. Applying our proposed algorithm to the optimal switching control model, we obtain the optimal number of switchings and the optimal feeding strategy of glycerol. The optimal number of switchings is 28. Moreover, the optimal feeding rates of glycerol during the total 14 feeding processes are shown in Fig. 8.1. The optimal time intervals for the batch processes are also listed in the Table 8.3.

Under this optimal control strategy in the fed-batch fermentation process, we get the concentration of 1,3-PD at the terminal time that is 975.319 mmol $L^{-1}$, which is increased by 22.34 % in comparison with 797.23 mmol $L^{-1}$ in the experiment. Figure 8.2 presents the concentration changes of glycerol and 1,3-PD, respectively. From Fig. 8.2, we can see that the concentration of glycerol maintains in the range of $\alpha_*$ and $\alpha^*$ after the first batch process. It is important for microorganism growth to effectively avoid the inhibition of excessive substrate. In addition, this is also the nature of our modeling the fed-batch fermentation process as an optimal switching control problem in this chapter. Figure 8.2 also shows that the concentration of 1,3-PD at the switchings from batch process to feeding process is firstly decrease and then increase. This is due to the fact that the dilute effect of the feed of glycerol and alkali excesses the production of 1,3-PD in the early stage of the feed and the latter surpasses the former in the following stage of the feed. Furthermore, Fig. 8.2 confirms that the concentration of 1,3-PD at the terminal time can be increased considerably. In addition, from Fig. 8.2 we can see that the volume change of culture fluid in the fermenter verifies the batch and feeding processes in the fed-batch fermentation process and the given volume $\Delta$ is also attained at the terminal time.

## 8.6 Conclusion

In this chapter, a state-based switched system was proposed to formulate the microbial fed-batch fermentation. This system is based on two facts in the fermentation process: (i) the fermentation process switches between the batch process and the feeding process, and (ii) the feeding of glycerol should be kept in a given range such that it can not only provide sufficient nutrition for cells growth but also can effectively avoid the inhibitory effect of excessive glycerol on the cells growth. To maximize the concentration of 1,3-PD at the terminal time, an optimal switching control problem was presented. To develop an efficient numerical computational method, we reformulated the optimal switching control problem as a two-level optimization problem. Numerical results show that the validity of the mathematical model and the effectiveness of the solution method.

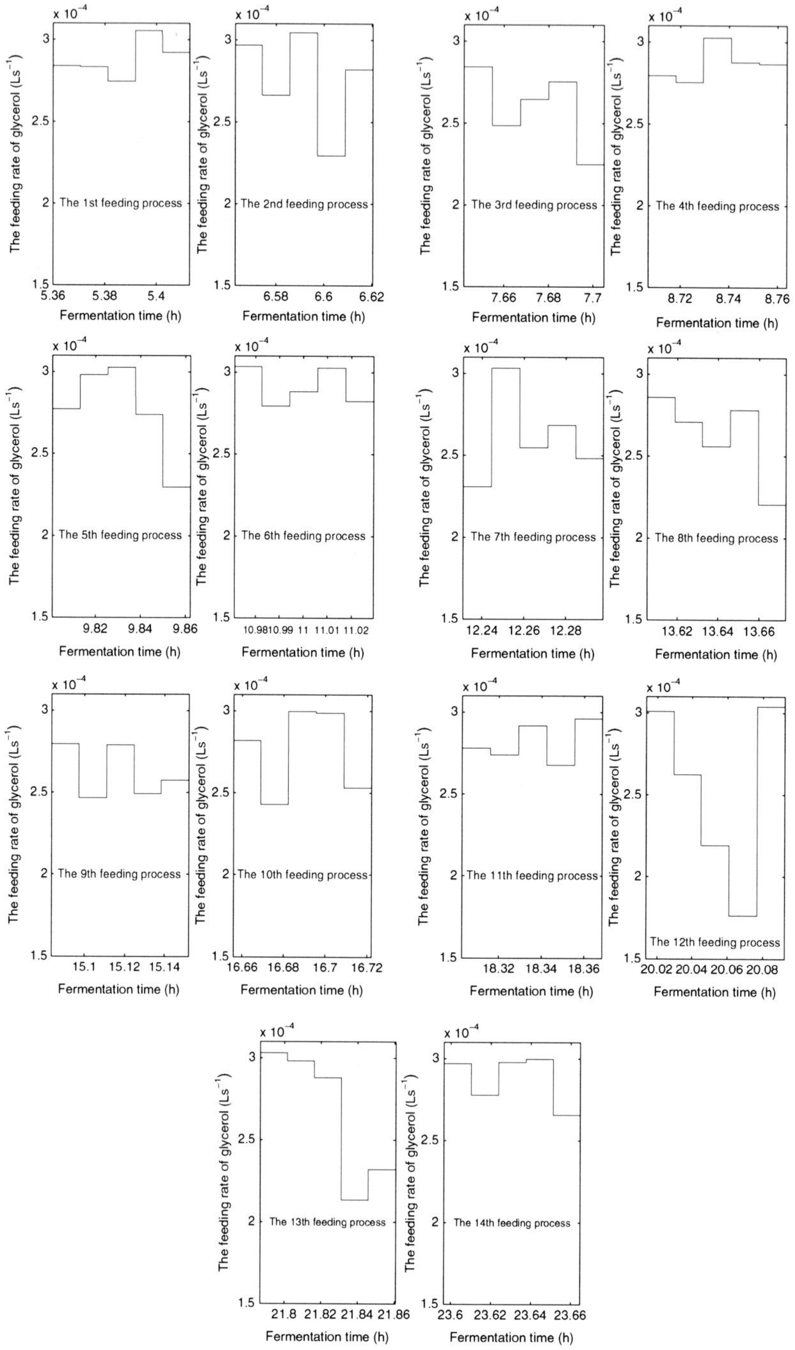


**Fig. 8.1** The optimal feeding rates of glycerol in the feeding processes of fed-batch fermentation process

**Table 8.3** The optimal time intervals of the batch processes in fed-batch fermentation process

| Processes | Time intervals | Processes | Time intervals |
|---|---|---|---|
| The 1st batch process | [0, 5.3600] | The 9th batch process | [13.6731, 15.0836] |
| The 2nd batch process | [5.4129, 6.5627] | The 10th batch process | [15.1520, 16.6557] |
| The 3rd batch process | [6.6204, 7.6423] | The 11th batch process | [16.7221, 18.3029] |
| The 4th batch process | [7.7051, 8.7066] | The 12th batch process | [18.3686, 20.0137] |
| The 5th batch process | [8.7638, 9.8010] | The 13th batch process | [20.0925, 21.7867] |
| The 6th batch process | [9.8622, 10.9710] | The 14th batch process | [21.8606, 23.5966] |
| The 7th batch process | [11.0292, 12.2311] | The 15th batch process | [23.6647, 24.1600] |
| The 8th batch process | [12.2980, 13.6052] | | |

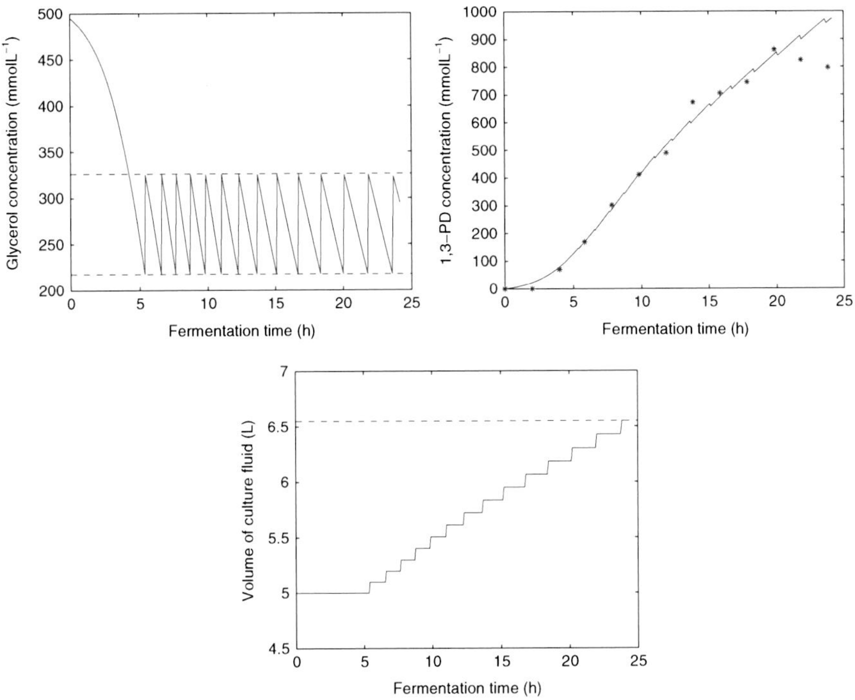


**Fig. 8.2** The concentration changes of glycerol, and 1,3-PD and the volume changes of culture fluid in the fed-batch fermentation process

# Chapter 9
# Optimal Parameter Selection of Multistage Time-Delay Systems

## 9.1 Introduction

In this chapter, we consider optimal parameter selection of a multistage time-delay system arising in fed-batch fermentation. Time delays exist in the process of glycerol bioconversion to 1,3-PD [172, 270]. Several reasons may be responsible for the occurrence of the delays in the fermentation process: A cell has to undergo some change or growth process for which it needs some time before it reacts with others; the substrate and the products have to be transported across the cell membrane requiring a certain amount of time for transport; sometimes, either because of lack of knowledge or in order to reduce complexity, it is appropriate to omit a number of intermediate steps in the reaction system for which the processing time is not negligible and has to be implemented as a delay [2, 164]. Thus, time delays have to be incorporated into mathematical models in formulating the fermentation process.

In this chapter, we propose a multistage time-delay system to formulate the fermentation process. Then, due to the effect of time delay and the high number of the kinetic parameters in the system, parametric sensitivity analysis is used to determine the key parameters. Parametric sensitivity analysis, i.e., the study of the influence of the parameters of a model on its solution, plays an important role in design, modeling, and parameter identification [75, 194]. In particular, the sensitivity analysis of time-delay systems had been investigated in the literature; see, for example, [10, 15, 124, 209]. Nevertheless, calculating the sensitivity functions is a very difficult task.

In this chapter, by solving the sensitivity functions numerically using the auxiliary system method, the key parameters are obtained. On this basis, an optimal parameter selection model involving the nonlinear time-delay system is presented, and the improved particle swarm optimization (PSO) algorithm in Chap. 5 is to seek



C. Liu, Z. Gong, *Optimal Control of Switched Systems Arising in Fermentation Processes*, Springer Optimization and Its Applications 97,
DOI 10.1007/978-3-662-43793-3_9

the optimal key parameters. Numerical results show that the multistage time-delay system can describe the fed-batch fermentation process better than the results previously reported.

The main reference of this chapter is [145].

## 9.2 Problem Formulation

### 9.2.1 Multistage Time-Delay Systems

During the fermentation process, the production of new biomass is delayed by the amount of time it takes to metabolize the nutrients. Thus, it is necessary to include time delays for the biomass formation in modeling the fermentation process. According to the fermentation process, we assume that

**Assumption 9.1.** *Biomass, substrate, 1,3-PD, acetate, and ethanol concentrations in reactor at time $t$ are determined by biomass concentration at time $t-\alpha$.*

Under Assumptions 5.2 and 9.1, the fed-batch fermentation process can be formulated as the following multistage time-delay system:

$$\begin{cases} \dot{\boldsymbol{x}}(t) = \boldsymbol{f}^i(t, \boldsymbol{x}(t), \boldsymbol{x}(t-\alpha), \boldsymbol{p}), \\ \boldsymbol{x}(t_{i-1}+) = \boldsymbol{x}(t_{i-1}),\ t \in (t_{i-1}, t_i],\ i = 1, 2, \ldots, 2N+1, \\ \boldsymbol{x}(t_0) = \boldsymbol{x}_0, \\ \boldsymbol{x}(t) = \boldsymbol{\phi}(t),\ t \in [-\tilde{\alpha}, t_0], \end{cases} \tag{9.1}$$

where $\boldsymbol{x}(t) := (x_1(t), x_2(t), x_3(t), x_4(t), x_5(t))^\top \in \mathbb{R}^5_+$ is the system state vector whose components represent the extracellular concentrations of biomass, glycerol, 1,3-PD, acetic acid, and ethanol at time $t$ in the fermentor, respectively. $t_i, i \in \Lambda := \{1, 2, \ldots, 2N+1\}$, is the switching instant such that $0 = t_0, t_{i-1} < t_i, i \in \Lambda$, and $t_{2N+1} = T$, which is decided a priori in the experiment. In particular, $t_{2j+1}$ is the moment of adding glycerol, at which the fermentation process switches to a feeding process, and $t_{2j+2}$ denotes the moment of ending the flow of glycerol, at which the fermentation process switches to a batch process, $j \in \bar{\Lambda}_1 := \{0, 1, 2, \ldots, N-1\}$. $T$ is the terminal time of the fermentation, $\tilde{\alpha} > 0$ is a given real number, $\boldsymbol{x}_0$ is a given initial state, and $\boldsymbol{\phi}(t) \in C^1\left([-\tilde{\alpha}, 0], \mathbb{R}^5_+\right)$ is a given initial function in which $C^1\left([-\tilde{\alpha}, 0], \mathbb{R}^5_+\right)$ is the Banach space of continuously differentiable functions mapping the interval $[-\tilde{\alpha}, 0]$ into $\mathbb{R}^5$. Furthermore, for $t \in \left(t_{2j}, t_{2j+1}\right], j \in \bar{\Lambda}_2 := \{0, 1, \ldots, N\}$,

$$f_\ell^{2j+1}(t, \boldsymbol{x}(t), \boldsymbol{x}(t-\alpha), \boldsymbol{p}) = q_\ell(\boldsymbol{x}(t)) x_1(t-\alpha), \quad \ell = 1, 3, 4, 5, \tag{9.2}$$

and

$$f_2^{2j+1}(t, \boldsymbol{x}(t), \boldsymbol{x}(t-\alpha), \boldsymbol{p}) = -q_2(\boldsymbol{x}(t)) x_1(t-\alpha); \tag{9.3}$$

for $t \in \left(t_{2j+1}, t_{2j+2}\right], j \in \bar{\Lambda}_1$,

$$f_\ell^{2j+2}(t, \boldsymbol{x}(t), \boldsymbol{x}(t-\alpha), \boldsymbol{p}) = q_\ell(\boldsymbol{x}(t))x_1(t-\alpha) - D(t)x_\ell(t),\ \ell = 1, 3, 4, 5, \tag{9.4}$$

and

$$f_2^{2j+2}(t, \boldsymbol{x}(t), \boldsymbol{x}(t-\alpha), \boldsymbol{p}) = D(t)\left(\frac{c_{s0}}{1+r} - x_2(t)\right) - q_2(\boldsymbol{x}(t))x_1(t-\alpha). \tag{9.5}$$

In (9.4) and (9.5), $r$ is the velocity ratio of adding alkali to glycerol. $c_{s0}$ denotes the concentration of initial feed of glycerol in medium. $D(t)$ is the dilution rate at time $t$ defined as

$$D(t) = \frac{(1+r)v_i}{V(t)}, \tag{9.6}$$

$$V(t) = V_0 + \sum_{j=1}^{i-1}(1+r)v_j\left(t_j - t_{j-1}\right) + (1+r)v_i \cdot (t - t_{i-1}). \tag{9.7}$$

In (9.6) and (9.7), $v_i \geqslant 0$ is the feeding rate of glycerol in $(t_{i-1}, t_i], i \in \Lambda$, and $V_0$ is the initial volume of culture fluid in the fermentor. The specific growth rate of cells $q_1(\boldsymbol{x}(t))$, the specific consumption rate of substrate $q_2(\boldsymbol{x}(t))$, and the specific formation rates of products $q_\ell(\boldsymbol{x}(t)), \ell = 3, 4, 5$, are expressed as the following equations:

$$q_1(\boldsymbol{x}(t)) = \frac{\Delta_1 x_2(t)}{x_2(t) + k_1}\prod_{\ell=2}^{5}\left(1 - \frac{x_\ell(t)}{x_\ell^*}\right)^{n_\ell}, \tag{9.8}$$

$$q_2(\boldsymbol{x}(t)) = m_2 + q_1(\boldsymbol{x}(t))Y_2 + \frac{\Delta_2 x_2(t)}{x_2(t) + k_2}, \tag{9.9}$$

$$q_\ell(\boldsymbol{x}(t)) = -m_\ell + q_1(\boldsymbol{x}(t))Y_\ell + \frac{\Delta_\ell x_2(t)}{x_2(t) + k_\ell}, \quad \ell = 3, 4, \tag{9.10}$$

$$q_5(\boldsymbol{x}(t)) = q_2(\boldsymbol{x}(t))\left(\frac{c_1}{c_2 + q_1(\boldsymbol{x}(t))x_2(t)} + \frac{c_3}{c_4 + q_1(\boldsymbol{x}(t))x_2(t)}\right). \tag{9.11}$$

It should be noted that there exist critical concentrations of biomass, glycerol, 1,3-PD, acetate, and ethanol, outside which cells cease to grow. Thus, it is biologically meaningful to restrict the concentrations of biomass, glycerol, and products within a set $W$ defined as

$$\boldsymbol{x}(t) \in W := \prod_{\ell=1}^{5}\left[x_{*\ell}, x_\ell^*\right], \ \forall t \in [0, T], \tag{9.12}$$

where the critical concentrations for cell growth are as given in Chap. 5.

Due to the introduction of time delay in the mathematical model, the values of kinetic parameters in the system (9.1) may be different from the previous ones. Hence, the time delay $\alpha$ and the kinetic parameters, i.e.,

$$p := (\Delta_1, k_1, m_2, m_3, m_4, Y_2, Y_3, Y_4, \Delta_2, \Delta_3, \Delta_4, k_2, k_3, k_4, c_1, c_2, c_3, c_4)^\top \in \mathbb{R}^{18}$$

should be identified. Here, time delay $\alpha$ is assumed to be non-negative and bounded above by $\tilde{\alpha}$, that is,

$$\alpha \in \mathcal{D} := [0, \tilde{\alpha}] . \tag{9.13}$$

In addition, the admissible set of the kinetic parameter vectors is defined as

$$\mathcal{P} := \prod_{l=1}^{18} \left[\tilde{p}_{*l}, \tilde{p}_l^*\right], \tag{9.14}$$

where $\tilde{p}_{*l}$ and $\tilde{p}_l^*$ are the lower bound and the upper bound of the kinetic parameter $p_l$, respectively. The values of $\tilde{p}_{*l}$ and $\tilde{p}_l^*$ are obtained by decrements and increments of the kinetic parameters considered in [84, 90, 271].

### 9.2.2 *Properties of the Multistage Time-Delay Systems*

For the system (9.1), some important properties, e.g., the existence and uniqueness, boundedness, and differentiability of the solution, are discussed in the subsection.

**Theorem 9.1.** *The function* $f^i : (t_{i-1}, t_i] \times \mathbb{R}^5_+ \times \mathbb{R}^5_+ \times \mathbb{R}^{18} \to \mathbb{R}^5, i \in \Lambda$, *defined in* (9.2)–(9.5) *satisfies the following conditions:*

(a) $f^i$ *is continuous on* $(t_{i-1}, t_i]$ *for each* $(x, y, p) \in \mathbb{R}^5_+ \times \mathbb{R}^5_+ \times \mathbb{R}^{18}$ *and is continuously differentiable with respect to each of the components* $x$, $y$, *and* $p$ *for each* $t \in (t_{i-1}, t_i]$.
(b) *There exists a constant* $K > 0$ *such that*

$$\|f^i(t, x, y, p)\| \leqslant K(1 + \|x\| + \|y\|), \forall (t, x, y, p) \in (t_{i-1}, t_i] \times \mathbb{R}^5_+ \times \mathbb{R}^5_+ \times \mathcal{P},$$

*where* $\|\cdot\|$ *denotes the Euclidean norm.*

*Proof.* (a) This conclusion can be obtained by the expression of $f^i$ in (9.2)–(9.5).
(b) We can complete the proof using a similar method as that given for the proof of Proposition 5.1 in Chap. 5. □

**Theorem 9.2.** *For each* $(\alpha, p) \in \mathcal{D} \times \mathcal{P}$, *the system* (9.1) *has a unique continuous solution, denoted by* $x(\cdot|\alpha, p)$, *on* $[-\tilde{\alpha}, T]$. *Furthermore,* $x(\cdot|\alpha, p)$ *satisfies that*

$$x(t|\alpha, p) = x(t_{i-1}|\alpha, p) + \int_{t_{i-1}}^{t} f^i(s, x(s|\alpha, p), x(s-\alpha|\alpha, p), p)\mathrm{d}s,$$

$$\forall t \in (t_{i-1}, t_i], i \in \Lambda, \tag{9.15}$$

*and* $x(t|\alpha, p) = \phi(t), \forall t \in [-\tilde{\alpha}, 0]$.

*Proof.* The proof can be obtained by Theorem 9.1 and the theory of delay-differential equations [95]. □

**Theorem 9.3.** *Given the initial function* $\phi(t) \in C^1([-\tilde{\alpha}, 0], \mathbb{R}^5_+)$ *and for all* $(\alpha, p) \in \mathcal{D} \times \mathcal{P}$, *the unique solution* $x(\cdot|\alpha, p)$ *of the system* (9.1) *is uniformly bounded.*

*Proof.* Let $(\alpha, p) \in \mathcal{D} \times \mathcal{P}$. Then since $\phi(t)$ is continuous on $[-\tilde{\alpha}, 0]$, there exists a real number $0 \leqslant M' < +\infty$ such that

$$\sup\{\|\phi(t)\| \mid t \in [-\tilde{\alpha}, 0]\} \leqslant M',$$

Hence,

$$\|x(t|\alpha, p)\| \leqslant M', \forall t \in [-\tilde{\alpha}, 0]. \tag{9.16}$$

In view of Theorems 9.1 and 9.2, we obtain that

$$\begin{aligned}
\|x(t|\alpha, p)\| &\leqslant \|x_0\| + \sum_{j=1}^{i-1}\int_{t_{j-1}}^{t_j} \|f^j(s, x(s|\alpha, p), x(s-\alpha|\alpha, p), p)\|\mathrm{d}s \\
&\quad + \int_{t_{i-1}}^{t} \|f^i(s, x(s|\alpha, p), x(s-\alpha|\alpha, p), p)\|\mathrm{d}s \\
&\leqslant M' + K\int_0^t (1 + \|x(s|\alpha, p)\| + \|x(s-\alpha|\alpha, p)\|)\mathrm{d}s \\
&\leqslant M' + K\int_0^t (1 + \|x(s|\alpha, p)\|)\mathrm{d}s + K\int_{-\alpha}^{t} \|x(s|\alpha, p)\|\mathrm{d}s \\
&\leqslant M' + K\tilde{\alpha}M' + K\int_0^t (1 + 2\|x(s|\alpha, p)\|)\mathrm{d}s, \quad \forall t \in (0, T].
\end{aligned}$$

By Lemma 4.1, it follows that

$$\|x(t|\alpha, p)\| \leqslant \left(M' + K\tilde{\alpha}M' + KT\right)\exp(2KT), \quad \forall t \in (0, T].$$

Thus,

$$\|x(t|\alpha, p)\| \leqslant M, \quad \forall t \in [-\tilde{\alpha}, T],$$

where $M := \max\{M', (M' + K\tilde{\alpha}M' + KT)\exp(2KT)\}$. The proof is complete. □

*Remark 9.1.* For each $(\alpha, \boldsymbol{p}) \in \mathcal{D} \times \mathcal{P}$, the solution $\boldsymbol{x}(\cdot|\alpha, \boldsymbol{p})$ is a function of time. In detail, if time delay and the kinetic parameters are fixed, then the solution of the system (9.1) is a function defined on $[-\tilde{\alpha}, T]$. Alternatively, we can fix $t \in [-\tilde{\alpha}, T]$ and consider the function $\boldsymbol{x}(t|\cdot,\cdot) : \mathcal{D} \times \mathcal{P} \to \mathbb{R}^5_+$ that returns the value of the system state at time $t$ corresponding to a given pair in $\mathcal{D} \times \mathcal{P}$.

In view of theory of the delay-differential equations [95], the next theorem can be established.

**Theorem 9.4.** *For all $t \in [0, T]$, the function $\boldsymbol{x}(t|\cdot,\cdot)$ is continuous on $\mathcal{D} \times \mathcal{P}$. Moreover, $\boldsymbol{x}(t|\cdot,\cdot)$ is differentiable on $\mathcal{D} \times \mathcal{P}$ for each $t \in (t_{i-1}, t_i]$, $i \in \Lambda$.*

## 9.3 Parametric Sensitivity Analysis

Determining the time delay $\alpha$ is very unusual since the delay influences the system state implicitly through the system (9.1). In addition, identification of the kinetic parameters $\boldsymbol{p}$ is also a difficult problem because of the high number of the parameters. In this section, we shall use parametric sensitivity theory to establish the effect of time delay on system state and to select the key parameters to be identified in the system (9.1).

### 9.3.1 Sensitivity Functions

Sensitivity analysis studies how changes of a model output can be apportioned, qualitatively or quantitatively, to variations of the parameters. Parameters in the dynamical system exerting the most influence on the system state can be established through the sensitivity analysis. Those insensitive parameters, which are not obviously influential on the system state, perhaps can be set as constants in the sequent parameter identification process.

Based on Theorem 9.4 and [75], the sensitivity functions are now defined as the partial derivatives of the system state with respect to $\alpha$ and $\boldsymbol{p}$, i.e.,

$$S_\alpha(t) := \frac{\partial \boldsymbol{x}(t|\alpha, \boldsymbol{p})}{\partial \alpha}, \quad t \in [0, T], \tag{9.17}$$

and

$$S_{\boldsymbol{p}}(t) := \frac{\partial \boldsymbol{x}(t|\alpha, \boldsymbol{p})}{\partial \boldsymbol{p}}, \quad t \in [0, T], \tag{9.18}$$

respectively.

Calculating the sensitivity functions is a very difficult task; the auxiliary system method will be used to deduce the formulae of the sensitivity functions. The main reason is that the auxiliary time-delay systems can be solved simultaneously with the system (9.1). Define

$$\bar{\chi}(t) := \begin{cases} \dot{\boldsymbol{\phi}}(t), & \text{if } t \leqslant 0, \\ \boldsymbol{f}^i(t, \boldsymbol{x}(t), \boldsymbol{x}(t-\alpha), \boldsymbol{p}), & \text{if } t \in (t_{i-1}, t_i] \text{ for some } i \in \Lambda. \end{cases} \tag{9.19}$$

We first give the sensitivity function $S_\alpha(\cdot)$ in terms of the solution of an auxiliary time-delay system in the following theorem.

**Theorem 9.5.** *For each $t \in [0, T]$ and $\boldsymbol{p} \in \mathcal{P}$,*

$$S_\alpha(t) = \boldsymbol{\psi}(t|\alpha, \boldsymbol{p}), \quad \alpha \in \mathcal{D}, \tag{9.20}$$

*where $\boldsymbol{\psi}(\cdot|\alpha, \boldsymbol{p})$ is the solution of the following auxiliary delay-differential system:*

$$\begin{cases} \dot{\boldsymbol{\psi}}(t) = \dfrac{\partial \boldsymbol{f}^i(t, \boldsymbol{x}(t|\alpha,\boldsymbol{p}), \boldsymbol{x}(t-\alpha|\alpha,\boldsymbol{p}),\boldsymbol{p})}{\partial \boldsymbol{x}(t)} \boldsymbol{\psi}(t) + \dfrac{\partial \boldsymbol{f}^i(t, \boldsymbol{x}(t|\alpha,\boldsymbol{p}), \boldsymbol{x}(t-\alpha|\alpha,\boldsymbol{p}),\boldsymbol{p})}{\partial \boldsymbol{x}(t-\alpha)} \\ \qquad \times \boldsymbol{\psi}(t-\alpha) - \dfrac{\partial \boldsymbol{f}^i(t, \boldsymbol{x}(t|\alpha,\boldsymbol{p}), \boldsymbol{x}(t-\alpha|\alpha,\boldsymbol{p}),\boldsymbol{p})}{\partial \boldsymbol{x}(t-\alpha)} \bar{\chi}(t-\alpha), \forall t \in (t_{i-1}, t_i], \\ \boldsymbol{\psi}(t_{i-1}+) = \boldsymbol{\psi}(t_{i-1}), i \in \Lambda, \end{cases} \tag{9.21}$$

*with*

$$\boldsymbol{\psi}(t) = \boldsymbol{0}, \ t \in [-\tilde{\alpha}, 0]. \tag{9.22}$$

*Proof.* Let $\alpha \in \mathcal{D}$ be arbitrary but fixed. Define

$$\alpha + \epsilon$$

where $\epsilon$ is sufficiently small such that

$$\alpha + \epsilon \in [0, \tilde{\alpha}].$$

In view of Theorem 9.2, $\boldsymbol{x}(t|\alpha, \boldsymbol{p})$ and $\boldsymbol{x}(t|\alpha + \epsilon, \boldsymbol{p})$, $\forall t \in (t_{i-1}, t_i], i \in \Lambda$, can be written as

$$\begin{aligned} \boldsymbol{x}(t|\alpha, \boldsymbol{p}) = \boldsymbol{x}_0 &+ \sum_{j=1}^{i-1} \int_{t_{j-1}}^{t_j} \boldsymbol{f}^j(s, \boldsymbol{x}(s|\alpha, \boldsymbol{p}), \boldsymbol{x}(s-\alpha|\alpha, \boldsymbol{p}), \boldsymbol{p}) \mathrm{d}s \\ &+ \int_{t_{i-1}}^{t} \boldsymbol{f}^i(s, \boldsymbol{x}(s|\alpha, \boldsymbol{p}), \boldsymbol{x}(s-\alpha|\alpha, \boldsymbol{p}), \boldsymbol{p}) \mathrm{d}s, \end{aligned}$$

and

$$x(t|\alpha+\epsilon,p) = x_0 + \sum_{j=1}^{i-1}\int_{t_{j-1}}^{t_j} f^j(s, x(s|\alpha+\epsilon,p), x(s-\alpha-\epsilon|\alpha+\epsilon,p), p)\mathrm{d}s$$
$$+ \int_{t_{i-1}}^{t} f^i(s, x(s|\alpha+\epsilon,p), x(s-\alpha-\epsilon|\alpha+\epsilon,p), p)\mathrm{d}s,$$

respectively. Thus, it follows that for $t \in (t_{i-1}, t_i],\ i \in \Lambda$,

$$\begin{aligned}
\psi(t|\alpha,p) :&= \left.\frac{\mathrm{d}x(t|\alpha+\epsilon,p)}{d\epsilon}\right|_{\epsilon=0} = \frac{\partial x(t|\alpha,p)}{\partial\alpha} \\
&= \sum_{j=1}^{i-1}\int_{t_{j-1}}^{t_j}\left(\frac{\partial f^j(s,x(s|\alpha,p),x(s-\alpha|\alpha,p),p)}{\partial x(s)}\psi(s|\alpha,p)\right. \\
&\quad + \frac{\partial f^j(s,x(s|\alpha,p),x(s-\alpha|\alpha,p),p)}{\partial x(s-\alpha)}\psi(s-\alpha|\alpha,p) \\
&\quad \left. - \frac{\partial f^j(s,x(s|\alpha,p),x(s-\alpha|\alpha,p),p)}{\partial x(s-\alpha)}\bar{\chi}(s-\alpha)\right)\mathrm{d}s \\
&\quad + \int_{t_{i-1}}^{t}\left(\frac{\partial f^i(s,x(s|\alpha,p),x(s-\alpha|\alpha,p),p)}{\partial x(s)}\psi(s|\alpha,p)\right. \\
&\quad + \frac{\partial f^i(s,x(s|\alpha,p),x(s-\alpha|\alpha,p),p)}{\partial x(s-\alpha)}\psi(s-\alpha|\alpha,p) \\
&\quad \left. - \frac{\partial f^i(s,x(s|\alpha,p),x(s-\alpha|\alpha,p),p)}{\partial x(s-\alpha)}\bar{\chi}(s-\alpha)\right)\mathrm{d}s.
\end{aligned} \tag{9.23}$$

Furthermore, since the state vector $x$ is continuous on $[-\tilde{\alpha}, T]$ and the switching instant $t_{i-1}, i \in \Lambda$, is independent of the choice of the time delay $\alpha$, we have

$$\psi(t_{i-1}+) = \psi(t_{i-1}). \tag{9.24}$$

It is also clear that for $t \in [-\tilde{\alpha}, 0]$,

$$\psi(t|\alpha,p) := \left.\frac{\mathrm{d}\phi(t|\alpha,p)}{d\epsilon}\right|_{\epsilon=0} = \mathbf{0}. \tag{9.25}$$

Obviously, (9.23) and (9.24) in conjunction with (9.25) is the solution of time-delay system (9.21) and (9.22), thereby completing the proof. □

The next theorem gives the sensitivity function $S_p(\cdot)$ in terms of the solution of another auxiliary time-delay system.

**Theorem 9.6.** *For each $t \in [0, T]$ and $\alpha \in \mathcal{D}$,*

$$S_p(t) = \varphi(t|\alpha, p), \quad p \in \mathcal{P}, \tag{9.26}$$

*where $\varphi(\cdot|\alpha, p)$ is the solution of the following auxiliary delay-differential system*

$$\begin{cases} \dot{\varphi}(t) = \dfrac{\partial f^i(t, x(t|\alpha, p), x(t-\alpha|\alpha, p), p)}{\partial x(t)}\varphi(t) \\ \qquad + \dfrac{\partial f^i(t, x(t|\alpha, p), x(t-\alpha|\alpha, p), p)}{\partial x(t-\alpha)} \\ \qquad \times \varphi(t-\alpha) + \dfrac{\partial f^i(t, x(t|\alpha, p), x(t-\alpha|\alpha, p), p)}{\partial p}, \forall t \in (t_{i-1}, t_i], \\ \varphi(t_{i-1}+) = \varphi(t_{i-1}), i \in \Lambda, \end{cases} \tag{9.27}$$

*with*

$$\varphi(t) = \mathbf{0}, \ t \in [-\tilde{\alpha}, 0]. \tag{9.28}$$

*Proof.* The proof is similar to the proof that is given for Theorem 9.5. □

For comparison, the relative sensitivity functions will be used in the numerical simulations because they are nondimensional and allow for comparing the results for different parameters and states. These functions are defined as

$$\bar{S}^{\ell}_{\alpha}(t) := \frac{\alpha}{x_\ell(t|\alpha, p)} S^{\ell}_{\alpha}(t), \ \ell = 1, 2, \ldots, 5, \ t \in [0, T], \tag{9.29}$$

and

$$\bar{S}^{\ell}_{p_l}(t) := \frac{p_l}{x_\ell(t|\alpha, p)} S^{\ell}_{p_l}(t), \ l = 1, 2, \ldots, 18; \ \ell = 1, 2, \ldots, 5, \ t \in [0, T], \tag{9.30}$$

respectively. However, for values of $x_\ell(t|\alpha, p)$ close to zero, a very large relative sensitivity may be obtained due to the division by $x_\ell(t|\alpha, p)$. Therefore, in this chapter, the relative sensitivity values were set to zero for all state values below 0.001.

The relative sensitivity functions (9.29) and (9.30) can in principle be obtained by Theorems 9.5 and 9.6, respectively. It should, however, be noted that the involving time-delay systems are highly nonlinear. Therefore, it is impossible to obtain analytical solutions of the above systems, and one has to resort to numerical simulations.

**Table 9.1** Parameters in computing the solution to the system (9.1)

| Parameters | Values |
|---|---|
| $\boldsymbol{x}_0$ | $(0.1115\,\mathrm{g\,L^{-1}}, 495\,\mathrm{mmol\,L^{-1}}, 0, 0, 0)^{\top}$ |
| $r$ | 0.75 |
| $c_{s0}$ | 10,762 $\mathrm{mmol\,L^{-1}}$ |
| $N$ | 783 |
| $V_0$ | 5 L |
| $\tilde{\alpha}$ | 12 h |
| $T$ | 27.1 h |

**Table 9.2** The feeding rates of glycerol in Phs. I–XI

| Phases | I–II,IV–V | III | VI | VII | VIII–XI |
|---|---|---|---|---|---|
| Feeding rates ($\mathrm{mL\,s^{-1}}$) | 0.2103 | 0.1992 | 0.2214 | 0.2437 | 0.2548 |

### 9.3.2 *Numerical Simulation Results*

The parametric sensitivity analysis for concentrations of biomass, glycerol, 1,3-PD, acetic acid, and ethanol with respect to time delay and kinetic parameters is investigated. The solution of the system (9.1) as well as the sensitivity functions in Theorems 9.5 and 9.6 were solved numerically using Matlab 7.10.0 (The Mathworks Inc.) and the intrinsic delay-differential equations (DDEs) with constant delay solver DDE23 which solved DDEs using explicit Runge–Kutta triples [225]. In particular, the relative error tolerance and the absolute error tolerance were set as $10^{-6}$ and $10^{-8}$, respectively. The cubic spline interpolation method [234] was adopted to construct the fitting curves before zero time point such that they pass through all the experimental data. The parameters needed in computing the solution to the system (9.1) are listed in Table 9.1. The feeding process began at $t_1 = 5.33$ h.The start of the feeding moment $t_{2j+1}$ and the stop of feeding moment $t_{2j+2}$, $j \in \Lambda_3 := \{0, 1, \ldots, 782\}$, were determined by the experiment. In the computational process, the fermentation process was partitioned into the first batch phase (Bat. Ph.) and phases I–XI (Phs. I–XI) according to the actual experiment. In each one of Phs. I–XI, the same feeding strategy was applied. The feeding rates of glycerol in Phs. I–XI are listed in Table 9.2. The value $\alpha = 0.25$ h for the time delay and the value $\boldsymbol{p} = (0.876, 0.28, 0.5953, 4.9029, 0.97, 128.205, 67.69, 33.07, 8.7388, 11.89089, 5.74, 17.7296, 15.50, 85.71, 0.025, 0.06, 5.18, 50.45)^{\top}$ for the kinetic parameter vector were taken as the initial values of time delay and the kinetic parameters. These values are obtained from previous work [14, 90]. The durations of the feeding processes in Phs. I–XI were 5, 7, 8, 7, 6, 4, 3, 2, 1, 2, and 1 s in each 100 s, leaving 95, 93, 92, 93, 94, 96, 97, 98, 99, 98, and 99 s for batch cultures, respectively. It should be mentioned that this approach had been adopted to obtain the experimental data and to identify the parameters of nonlinear multistage system in [90]. Furthermore, by Eqs. (9.29) and (9.30), the relative sensitivity functions $\bar{S}^{\ell}_{\alpha}(t)$ and $\bar{S}^{\ell}_{p_l}(t)$, $l = 1, 2, \ldots, 18$; $\ell = 1, 2, \ldots, 5$, were also computed. For comparison, the relative sensitivities are plotted against time. Figures 9.1–9.5 show

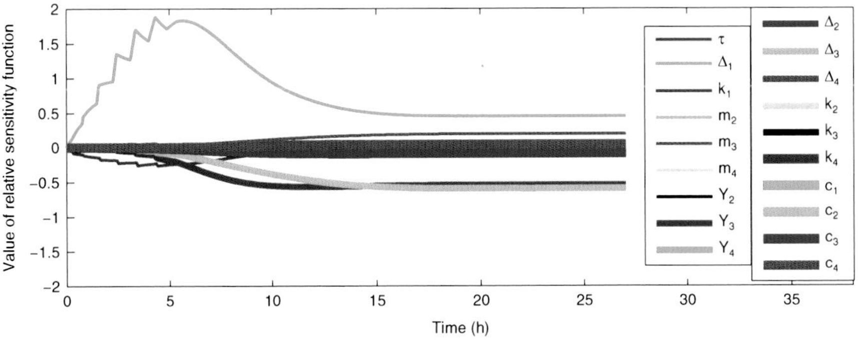


**Fig. 9.1** The relative sensitivities of biomass concentration with respect to delay and kinetic parameters

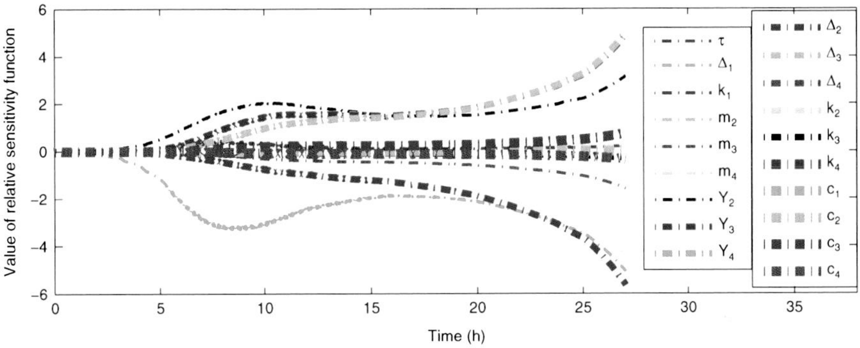


**Fig. 9.2** The relative sensitivities of glycerol concentration with respect to delay and kinetic parameters

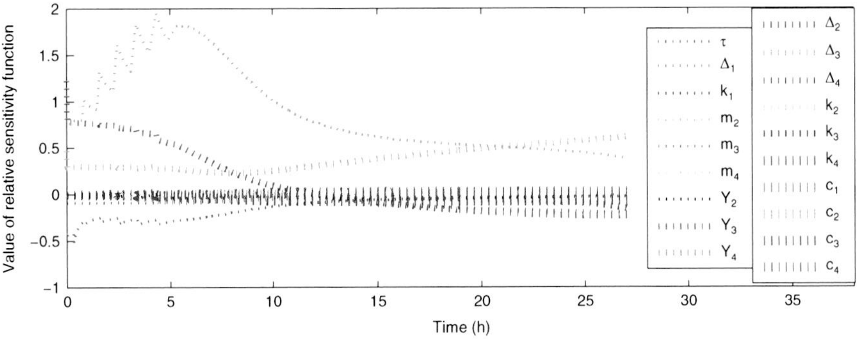


**Fig. 9.3** The relative sensitivities of 1,3-PD concentration with respect to delay and kinetic parameters

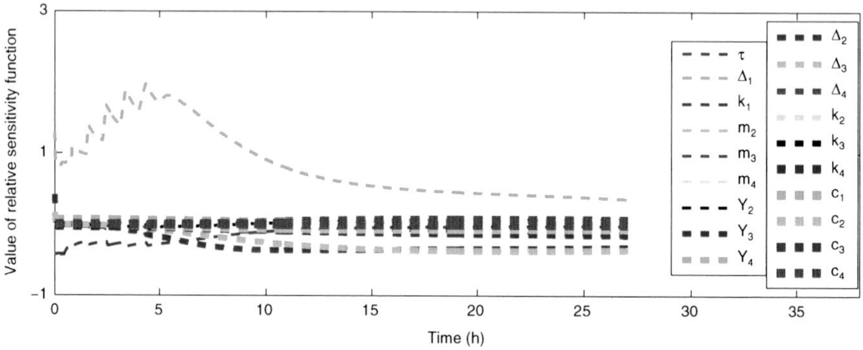


**Fig. 9.4** The relative sensitivities of acetic acid concentration with respect to delay and kinetic parameters

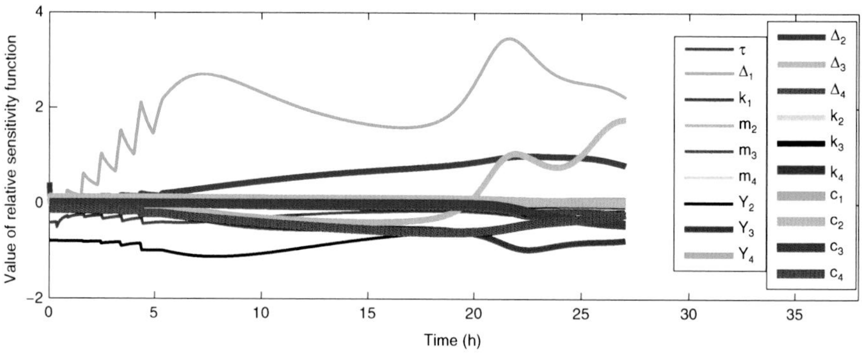


**Fig. 9.5** The relative sensitivities of ethanol concentration with respect to delay and kinetic parameters

the relative sensitivity curves for the concentrations of biomass, glycerol, 1,3-PD, acetic acid, and ethanol, respectively. It can be observed from Figs. 9.1 to 9.5 that some curves among the 19 ones stay near zero, that is, the effect of these parameters on the concentration changes can be assumed to be very diminutive, and these parameters can thus be concluded to have low sensitivities. In detail, we assume that if the maximal absolute value of specific sensitivity curve is less than 0.3, the parameter has low sensitivity. Table 9.3 lists the parameters of high sensitivity for space convenience. The check marks in Table 9.3 imply high sensitivities of parameters to certain $\boldsymbol{x}$ component.

Let the parameters that have high sensitivities in at least one of Figs. 9.1–9.5 be key parameters. Specifically, they are $\alpha$, $\Delta_1, m_2, m_3, Y_2, Y_3, \Delta_2, \Delta_3, c_3$, and $c_4$. For low-sensitivity parameters, substituting values in [90] should be adequate. More correct and reliable values are necessary only for high-sensitivity parameters. As a

**Table 9.3** The key parameters in the system (9.1)

| | $\alpha$ | $\Delta_1$ | $m_2$ | $m_3$ | $Y_2$ | $Y_3$ | $\Delta_2$ | $\Delta_3$ | $c_3$ | $c_4$ |
|---|---|---|---|---|---|---|---|---|---|---|
| $x_1$ | | √ | | | | √ | | √ | | |
| $x_2$ | √ | √ | √ | √ | √ | √ | √ | √ | √ | √ |
| $x_3$ | √ | √ | | | | √ | | √ | | |
| $x_4$ | √ | √ | | | | √ | | √ | | |
| $x_5$ | √ | √ | | √ | √ | √ | √ | √ | | √ |

result, the key parameters to be identified are simplified to time delay $\alpha$ and $\bar{\boldsymbol{p}} := (\Delta_1, m_2, m_3, Y_2, Y_3, \Delta_2, \Delta_3, c_3, c_4)^\top$. Notice that the number of parameters to be identified reduces to 10 from the original 19.

## 9.4 Optimal Parameter Selection Problems

The optimal parameter selection problem for a time-delay system is generally to adjust values of time delay and the parameters so that the discrepancy between predicted and observed system output is as small as possible [44–46, 159]. An optimal parameter selection problem may be resolved by fitting parameterized solutions to experimental data through minimizing a least-squares objective function. In this section, the key parameters obtained in the previous section will be taken as the parameters to be identified, and the other parameters in $\boldsymbol{p}$ are regarded as constants whose values take from [90].

### *9.4.1 Optimal Parameter Selection Models*

In the fed-batch fermentation process, we have measured $n$ experimental data. However, since the by-products of acetic acid and ethanol can change the pH values, alkali is fed into the fermentor to maintain a suitable environment for cell growth. The measured concentrations of acetic acid and ethanol are inaccurate due to this feed. As a result, the experimental concentrations of biomass, glycerol, and 1,3-PD are only used to identify the key parameters. First of all, we denote the concentrations of biomass, glycerol, and 1,3-PD measured at the moment $t_\imath$ in the experiment by $y_1^\imath, y_2^\imath$, and $y_3^\imath$, $\imath \in \{1, 2, \ldots, n\}$, respectively. Furthermore, let $\boldsymbol{x}(\cdot|\alpha, \bar{\boldsymbol{p}})$ be the solution of the system (9.1) corresponding to a pair $(\alpha, \bar{\boldsymbol{p}}) \in \mathcal{D} \times \bar{\mathcal{P}}$, where $\bar{\mathcal{P}}$ is the corresponding range for the key parameter vector $\bar{\boldsymbol{p}}$ in $\mathcal{P}$. Now, we consider the following cost function [90]

$$J(\alpha, \bar{\boldsymbol{p}}) = \sum_{\ell=1}^{3} \sum_{\imath=1}^{n} (x_\ell(t_\imath|\alpha, \bar{\boldsymbol{p}}) - y_\ell^\imath)^2 \tag{9.31}$$

to evaluate the errors between the computational values $\boldsymbol{x}(t_\imath|\alpha, \bar{\boldsymbol{p}})$ and the experimental data $\boldsymbol{y}^\imath$ at the moment $t_\imath$.

Given the system (9.1), our purpose is to find a $(\alpha, \bar{\boldsymbol{p}}) \in \mathcal{D} \times \bar{\mathcal{P}}$ such that the constraint (9.12) is satisfied and the cost function (9.31) is minimized. Hence, the optimal parameter selection model can be stated formally as

$$\begin{aligned} \text{(OPSM)} \qquad & \min\ J(\alpha, \bar{\boldsymbol{p}}) \\ & \text{s.t. } \boldsymbol{x}(t|\alpha, \bar{\boldsymbol{p}}) \in W,\ t \in [0, T], \\ & \quad (\alpha, \bar{\boldsymbol{p}}) \in \mathcal{D} \times \bar{\mathcal{P}}. \end{aligned}$$

The existence of the optimal solution for the optimal parameter selection model (OPSM) can be established as follows.

**Theorem 9.7.** (OPSM) *has an optimal solution, that is, there exists* $(\alpha^*, \bar{\boldsymbol{p}}^*) \in \mathcal{D} \times \bar{\mathcal{P}}$ *such that*

$$J(\alpha^*, \bar{\boldsymbol{p}}^*) \leqslant J(\alpha, \bar{\boldsymbol{p}}),\ \forall (\alpha, \bar{\boldsymbol{p}}) \in \mathcal{D} \times \bar{\mathcal{P}}. \tag{9.32}$$

*Proof.* Define the admissible set of the key parameters as

$$\begin{aligned} \mathcal{F} := \{(\alpha, \bar{\boldsymbol{p}})|\ & \boldsymbol{x}(\cdot|\alpha, \bar{\boldsymbol{p}}) \text{ is the solution of the system (9.1) on } [-\tilde{\alpha}, T] \\ & \text{and } \boldsymbol{x}(t|\alpha, \bar{\boldsymbol{p}}) \in W \text{ for } t \in [0, T]\}. \end{aligned} \tag{9.33}$$

Obviously, $\mathcal{F}$ is nonempty. Moreover, $\mathcal{F} \subseteq \mathcal{D} \times \bar{\mathcal{P}}$ is a bounded set due to the compactness of the set $\mathcal{D} \times \bar{\mathcal{P}}$. Then, for any sequence $\{(\alpha^i, \bar{\boldsymbol{p}}^i)\}_{i=1}^{\infty} \subseteq \mathcal{F}$, there exists at least one subsequence $\left\{\left(\hat{\alpha}^{i_j}, \hat{\bar{\boldsymbol{p}}}^{i_j}\right)\right\} \subseteq \{(\alpha^i, \bar{\boldsymbol{p}}^i)\}$ such that $\left(\hat{\alpha}^{i_j}, \hat{\bar{\boldsymbol{p}}}^{i_j}\right) \to \left(\hat{\alpha}, \hat{\bar{\boldsymbol{p}}}\right)$ as $j \to \infty$. Now, for each $j$, suppose $\boldsymbol{x}\left(\cdot|\hat{\alpha}^{i_j}, \hat{\bar{\boldsymbol{p}}}^{i_j}\right)$ is the solution of the system (9.1) and $\boldsymbol{x}\left(t|\hat{\alpha}^{i_j}, \hat{\bar{\boldsymbol{p}}}^{i_j}\right) \in W$ for all $t \in [0, T]$, then $\boldsymbol{x}\left(\cdot|\hat{\alpha}, \hat{\bar{\boldsymbol{p}}}\right)$ is also a solution of the system (9.1) and $\boldsymbol{x}\left(\cdot|\hat{\alpha}, \hat{\bar{\boldsymbol{p}}}\right) \in W$ in view of Theorem 9.4 and the compactness of the $W$. Namely, $\left(\hat{\alpha}, \hat{\bar{\boldsymbol{p}}}\right) \in \mathcal{F}$, which implies the closeness of the set $\mathcal{F}$. Furthermore, since the cost function $J(\alpha, \bar{\boldsymbol{p}})$ is continuous on $\mathcal{D} \times \bar{\mathcal{P}}$, we confirm that (OPSM) has an optimal pair $(\alpha^*, \bar{\boldsymbol{p}}^*)$ such that (9.32) holds. This completes the proof. □

### 9.4.2 A Computational Procedure

(OPSM) is in essence a parameter optimization problem. However, since the constraint (9.12) in (OPSM) is a continuous state inequality constraint, we shall use the method in Chap. 6 to deal with these continuous inequality constraints.

To begin with, let

$$g_\ell(\boldsymbol{x}(t|\alpha, \bar{\boldsymbol{p}})) = x_\ell^* - x_\ell(t|\alpha, \bar{\boldsymbol{p}}),$$
$$g_{5+\ell}(\boldsymbol{x}(t|\alpha, \bar{\boldsymbol{p}})) = x_\ell(t|\alpha, \bar{\boldsymbol{p}}) - x_{*\ell},\ \ell = 1, 2, \ldots, 5.$$

The condition $\boldsymbol{x}(t|\alpha, \bar{\boldsymbol{p}}) \in W$ is equivalently transcribed into

$$G(\alpha, \bar{\boldsymbol{p}}) = 0, \tag{9.34}$$

where $G(\alpha, \bar{\boldsymbol{p}}) := \sum_{\iota=1}^{10} \int_0^T \min\{0, g_\iota(\boldsymbol{x}(t|\alpha, \bar{\boldsymbol{p}}))\}\mathrm{d}t$. However, the equality constraint (9.34) is non-differentiable at the points when $g_\iota = 0$. We replace (9.34) with

$$\tilde{G}_{\varepsilon,\gamma}(\alpha, \bar{\boldsymbol{p}}) := \gamma + \sum_{\iota=1}^{10} \int_0^T \pi_\varepsilon(g_\iota(\boldsymbol{x}(s|\alpha, \bar{\boldsymbol{p}})))\mathrm{d}s \geqslant 0, \tag{9.35}$$

where $\varepsilon > 0$, $\gamma > 0$ and

$$\pi_\varepsilon(\eta) = \begin{cases} \eta, & \text{if } \eta < -\varepsilon, \\ -\dfrac{(\eta - \varepsilon)^2}{4\varepsilon}, & \text{if } -\varepsilon \leqslant \eta \leqslant \varepsilon, \\ 0, & \text{if } \eta > \varepsilon. \end{cases} \tag{9.36}$$

Thus, (OPSM) is approximated by a sequence of nonlinear programming problems (OPSM$_{\varepsilon,\gamma}$) defined by replacing constraint (9.34) with (9.35). Clearly, for each $\varepsilon$ and $\gamma$, (OPSM$_{\varepsilon,\gamma}$) is a mathematical programming in canonical form.

The next theorem shows that for any $\varepsilon > 0$, if $\gamma$ is chosen sufficiently small, the solution of the corresponding problem (OPSM$_{\varepsilon,\gamma}$) will satisfy the continuous state inequality constraint (9.12).

**Theorem 9.8.** *For each $\varepsilon > 0$, there exists a $\gamma(\varepsilon) > 0$ such that if* (9.35) *with $\gamma < \gamma(\varepsilon)$ is satisfied for some $(\alpha, \bar{\boldsymbol{p}}) \in \mathcal{D} \times \bar{\mathcal{P}}$, then the original constraint* (9.12) *is also satisfied at $(\alpha, \bar{\boldsymbol{p}}) \in \mathcal{D} \times \bar{\mathcal{P}}$.*

The proof of Theorem 9.8 can be found in Chapter 8 of [240]. On the basis of this theorem, (OPSM) can be solved through solving a sequence of problems $\{(\text{OPSM}_{\varepsilon,\gamma})\}$. In the computational process, the gradients of constraint $\tilde{G}_{\varepsilon,\gamma}(\alpha, \bar{\boldsymbol{p}})$ with respect to each key parameter are needed. However, the traditional methods for computing the gradient of the constraint $\tilde{G}_{\varepsilon,\gamma}(\alpha, \bar{\boldsymbol{p}})$ involve integrating two systems of differential equations—the state system and the costate system—successively in different directions, which is difficult to implement in computation process [157, 240]. In contrast, we will develop a new scheme for computing these gradients of the constraint $\tilde{G}_{\varepsilon,\gamma}(\alpha, \bar{\boldsymbol{p}})$ on the basis of Theorems 9.5 and 9.6.

**Theorem 9.9.** *For each $\varepsilon > 0$ and $\gamma > 0$, the gradient of the constraint $\tilde{G}_{\varepsilon,\gamma}(\alpha, \bar{\boldsymbol{p}})$ defined in* (9.35) *with respect to $\alpha$ is*

$$\frac{\partial \tilde{G}_{\varepsilon,\gamma}(\alpha, \bar{\boldsymbol{p}})}{\partial \alpha} = \sum_{\iota=1}^{10} \int_0^T \frac{\partial \pi_\varepsilon(g_\iota(\boldsymbol{x}(t|\alpha, \bar{\boldsymbol{p}})))}{\partial g_\iota} \frac{\partial g_\iota(\boldsymbol{x}(t|\alpha, \bar{\boldsymbol{p}}))}{\partial \boldsymbol{x}} \boldsymbol{\zeta}(t)\mathrm{d}t, \quad (9.37)$$

*where $\boldsymbol{\zeta}(\cdot)$ is the solution of the following time-delay system*

$$\begin{cases} \dot{\boldsymbol{\zeta}}(t) = \dfrac{\partial \boldsymbol{f}^i(t, \boldsymbol{x}(t|\alpha, \bar{\boldsymbol{p}}), \boldsymbol{x}(t-\alpha|\alpha, \bar{\boldsymbol{p}}), \bar{\boldsymbol{p}})}{\partial \boldsymbol{x}(t)} \boldsymbol{\zeta}(t) \\ \qquad + \dfrac{\partial \boldsymbol{f}^i(t, \boldsymbol{x}(t|\alpha, \bar{\boldsymbol{p}}), \boldsymbol{x}(t-\alpha|\alpha, \bar{\boldsymbol{p}}), \bar{\boldsymbol{p}})}{\partial \boldsymbol{x}(t-\alpha)} \\ \qquad \times \boldsymbol{\zeta}(t-\alpha) - \dfrac{\partial \boldsymbol{f}^i(t, \boldsymbol{x}(t|\alpha, \bar{\boldsymbol{p}}), \boldsymbol{x}(t-\alpha|\alpha, \bar{\boldsymbol{p}}), \bar{\boldsymbol{p}})}{\partial \boldsymbol{x}(t-\alpha)} \boldsymbol{\chi}(t-\alpha), \ \forall t \in (t_{i-1}, t_i], \\ \boldsymbol{\zeta}(t_{i-1}+) = \boldsymbol{\zeta}(t_{i-1}), \ i \in \Lambda, \end{cases} \quad (9.38)$$

*with*

$$\boldsymbol{\zeta}(t) = \boldsymbol{0}, \ t \in [-\tilde{\alpha}, 0]. \quad (9.39)$$

**Theorem 9.10.** *For each $\varepsilon > 0$ and $\gamma > 0$, the gradient of the constraint $\tilde{G}_{\varepsilon,\gamma}(\alpha, \bar{\boldsymbol{p}})$ defined in* (9.35) *with respect $\bar{\boldsymbol{p}}$ is*

$$\frac{\partial \tilde{G}_{\varepsilon,\gamma}(\alpha, \bar{\boldsymbol{p}})}{\partial \bar{\boldsymbol{p}}} = \sum_{\iota=1}^{10} \int_0^T \frac{\partial \pi_\varepsilon(g_\iota(\boldsymbol{x}(t|\alpha, \bar{\boldsymbol{p}})))}{\partial g_\iota} \frac{\partial g_\iota(\boldsymbol{x}(t|\alpha, \bar{\boldsymbol{p}}))}{\partial \boldsymbol{x}} \boldsymbol{\xi}(t)\mathrm{d}t, \quad (9.40)$$

*where $\boldsymbol{\xi}(\cdot)$ is the solution of the following time-delay system*

$$\begin{cases} \dot{\boldsymbol{\xi}}(t) = \dfrac{\partial \boldsymbol{f}^i(t, \boldsymbol{x}(t|\alpha, \bar{\boldsymbol{p}}), \boldsymbol{x}(t-\alpha|\alpha, \bar{\boldsymbol{p}}), \bar{\boldsymbol{p}})}{\partial \boldsymbol{x}(t)} \boldsymbol{\xi}(t) \\ \qquad + \dfrac{\partial \boldsymbol{f}^i(t, \boldsymbol{x}(t|\alpha, \bar{\boldsymbol{p}}), \boldsymbol{x}(t-\alpha|\alpha, \bar{\boldsymbol{p}}), \bar{\boldsymbol{p}})}{\partial \boldsymbol{x}(t-\alpha)} \\ \qquad \times \boldsymbol{\xi}(t-\alpha) + \dfrac{\partial \boldsymbol{f}^i(t, \boldsymbol{x}(t|\alpha, \bar{\boldsymbol{p}}), \boldsymbol{x}(t-\alpha|\alpha, \bar{\boldsymbol{p}}), \bar{\boldsymbol{p}})}{\partial \bar{\boldsymbol{p}}}, \ \forall t \in (t_{i-1}, t_i], \\ \boldsymbol{\xi}(t_{i-1}+) = \boldsymbol{\xi}(t_{i-1}), \ i \in \Lambda, \end{cases} \quad (9.41)$$

*with*

$$\boldsymbol{\xi}(t) = \boldsymbol{0}, \ t \in [-\tilde{\alpha}, 0]. \quad (9.42)$$

Now, each of problems $\{(\text{OPSM}_{\varepsilon,\gamma})\}$ can be solved by gradient-based optimization methods [240]. Nevertheless, all those techniques are only designed to find local optimal solutions. The need of global optimization techniques to avoid the spurious solutions often found by traditional gradient-based local methods had been highlighted in [144, 171, 180]. Therefore, we will solve each problem (OPSM$_{\varepsilon,\gamma}$) using the improved PSO algorithm in Chap. 5. On the basis of Theorems 9.8–9.10 and the improved PSO, the following algorithm to solve the (OPSM) can be developed.

**Algorithm 9.1.**

**Step 1.** Set $\varepsilon > 0, \gamma > 0, \beta_1 > 0, \beta_2 > 0, \bar{\varepsilon} > 0$, and $\bar{\gamma} > 0$.
**Step 2.** Solve (OPSM$_{\varepsilon,\gamma}$) using the improved PSO algorithm to give $(\alpha^*_{\varepsilon,\gamma}, \bar{\boldsymbol{p}}^*_{\varepsilon,\gamma})$.
**Step 3.** Check feasibility of $g_\iota(\boldsymbol{x}(t|\alpha^*_{\varepsilon,\gamma}, \bar{\boldsymbol{p}}^*_{\varepsilon,\gamma}) \geqslant 0$ for $t \in [0, T]$ and $\iota = 1, 2, \ldots, 10$.
**Step 4.** If $(\alpha^*_{\varepsilon,\gamma}, \bar{\boldsymbol{p}}^*_{\varepsilon,\gamma})$ is feasible, then go to Step 5. Otherwise, set $\gamma := \beta_1\gamma$. If $\gamma < \bar{\gamma}$, then go to Step 6. Otherwise, go to Step 2.
**Step 5.** Set $\varepsilon := \beta_2\varepsilon$. If $\varepsilon \geqslant \bar{\varepsilon}$, then go to Step 3. Otherwise, go to Step 6.
**Step 6.** Output $(\alpha^*_{\varepsilon,\gamma}, \bar{\boldsymbol{p}}^*_{\varepsilon,\gamma})$ and stop.

### *9.4.3 Numerical Results*

In the numerical computation, the medium composition cultivation conditions and the determination of biomass, substrate, and metabolites have been reported [48]. Algorithm 9.1 was applied to seeking the optimal key parameters in (OPSM) and all computations were implemented in Matlab 7.10.0. Here, the establishment of the initial function $\boldsymbol{\phi}(t)$, the parameters needed in solving the system (9.1), and the feeding rate of glycerol were the same as the ones used in Sect. 9.3.2. The start and stop moments of glycerol and alkali adding were determined by the experiment. The lower bounds and the upper bounds of the key parameter $\bar{\boldsymbol{p}}$ were $\bar{\boldsymbol{p}}_* = (0.438, 0.5, 2.45145, 0.0039, 33.845, 5.945445, 8.8648, 2.59, 25.225)^{\top}$ and $\bar{\boldsymbol{p}}^* = (1.314, 3.3, 7.35435, 0.0117, 101.535, 17.836335, 26.5944, 7.77, 75.675)^{\top}$, respectively [84, 90, 271]. In addition, the smoothing and feasible parameters were initially selected as $\varepsilon = 0.1$ and $\gamma = 0.01$. The parameters $\beta_1$ and $\beta_2$ were chosen as 0.1 and 0.01 until the solution obtained was feasible for the original problem. The process terminated when $\bar{\varepsilon} = 1.0 \times 10^{-8}$ and $\bar{\gamma} = 1.0 \times 10^{-7}$. In the improved PSO, the number of initial particles swarm $N^p$, the maximal iteration $M^p$, and the parameters $c_1^p, c_2^p, P_{cr}, M_1^p, M_2^p, \varepsilon_p$, which have the same meanings as those given in Chap. 5, were, respectively, 50, 200, 2, 2, 0.5, 100, 20, and $10^{-3}$.

By applying Algorithm 9.1, we obtained the optimal key parameter $\alpha^* = 0.4652$ h and $\bar{\boldsymbol{p}}^* = (0.8, 1.927, 3.2819, 0.0063, 80.6096, 6.8489, 10.3687, 2.81, 65.5226)^{\top}$. In particular, the strategy of dealing with state constraints by Theorem 9.9 was performed for eight times in the computation process. The performance of the system (9.1) is compared with that of the nonlinear multistage system which is an ordinary differential equation model in [90]. The relative errors $e_\ell$, $\ell = 1, 2, 3$, between the computational values and the experimental data for the two models were listed in Table 9.4 where the relative errors are defined as

**Table 9.4** The relative errors between the computational values and the experimental data

| Relative errors | $e_1$ (%) | $e_2$ (%) | $e_3$ (%) |
|---|---|---|---|
| Multistage time-delay system | 3.989 | 5.551 | 4.03 |
| Nonlinear multistage system [90] | 7.46 | 11.46 | 5.35 |

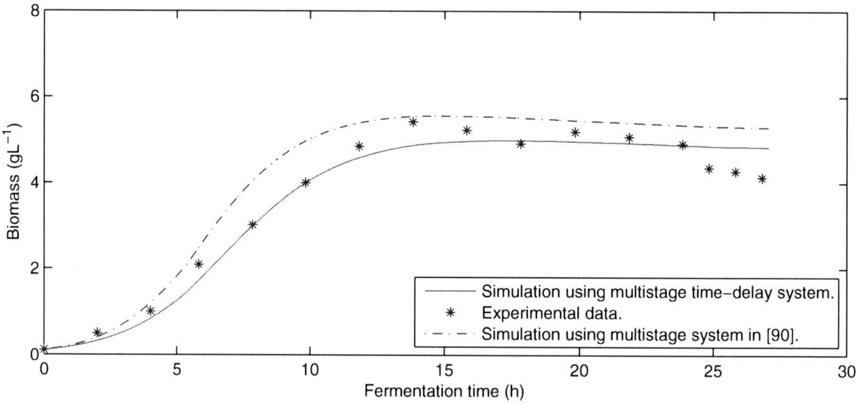


**Fig. 9.6** The concentration changes of biomass with respect to fermentation time

$$e_\ell = \frac{\sum_{\iota=1}^{n} |x_\ell(t_\iota|\alpha^*, \bar{p}^*) - y_\ell^\iota|}{\sum_{\iota=1}^{n} y_\ell^\iota}, \tag{9.43}$$

where $x_\ell(t_\iota|\alpha^*, \bar{p}^*)$ is the $\ell$th component of the solution to the system (9.1) under the optimal key parameters. From Table 9.4, we conclude that the relative errors are decreased compared with the ones in the nonlinear multistage system [90]. In particular, the concentration changes of biomass, glycerol, and 1,3-PD with respect to the optimal key parameters for the system (9.1) were respectively shown in Figs. 9.6–9.8. The simulation for the nonlinear multistage system in [90] and the experimental data were also presented in these figures for comparison. As a result, the system (9.1) fits the experimental data better than the one previously reported.

Furthermore, under the optimal key parameters $\alpha^*$ and $\bar{p}^*$, we recalculated the system (9.1) and obtained the predictive concentrations of biomass, glycerol, 1,3-PD, acetic acid, and ethanol at time 27.83 h which are 4.533 g L$^{-1}$, 174.79, 1,011.86, 203.65, and 195.58 mmol L$^{-1}$, respectively. In comparison with the experimental concentrations of biomass, glycerol, 1,3-PD, acetic acid, and ethanol, i.e., 4.38 g L$^{-1}$, 186.85, 1,035.0, 182.5, and 174.35 mmol L$^{-1}$, we can see that the predictive concentrations of biomass, glycerol, and 1,3-PD are well consistent with the experimental data. Nevertheless, the predictive concentrations of acetic acid and

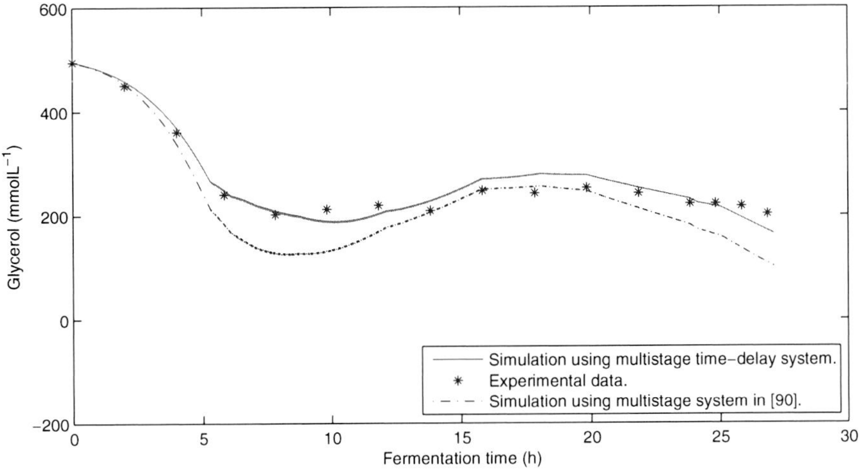


**Fig. 9.7** The concentration changes of glycerol with respect to fermentation time

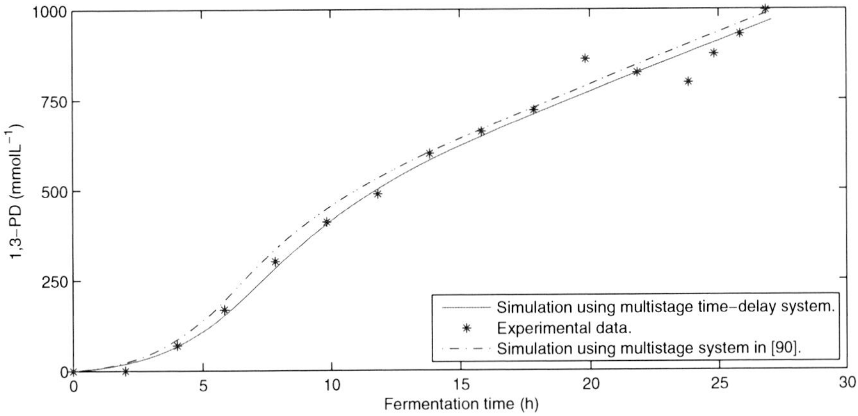


**Fig. 9.8** The concentration changes of 1,3-PD with respect to fermentation time

ethanol deviate from the experimental data about 11.59 and 12.18 %, respectively. The reasons why the deviations occurred might be that the experimental concentrations of acetic acid and ethanol were not incorporated in the parameter identification problem. This also verified the inaccuracies of the experimental concentrations of acetic acid and ethanol due to the feed of the alkali.

In all, from the numerical results, we can see that the proposed system introducing the time delay in modeling the fed-batch fermentation process is reasonable.

## 9.5 Conclusion

In this chapter, we investigated the optimal parameter selection problem of a multistage time-delay system arising in fed-batch fermentation. To determine the time-delay effect and decrease the number of the kinetic parameters, the parametric sensitivity analysis was investigated. On this basis, an optimal parameter selection model was presented and a computational procedure was developed to seek the optimal key parameters. Numerical results verified the validity of the mathematical model and the effectiveness of the computational method.

# Chapter 10
# Optimal Control of Multistage Time-Delay Systems

## 10.1 Introduction

In this chapter, we consider optimal control problem of a multistage time-delay system arising in fed-batch fermentation. As stated in Chap. 9, time delays exist in the process of glycerol bioconversion to 1,3-PD since a cell has to undergo some change or growth process for which it needs some time before it reacts with others. As a result, we propose a controlled multistage time-delay system, in which the flow rate of glycerol is taken as the control function and the terminal time as the optimization variable, to formulate the fed-batch process. The main goal of control the fermentation is to maximize the yield of 1,3-PD and reduce operation costs [153]. Thus, the mass of 1,3-PD per unit time is regarded as the performance index. By the way, many studies have considered the same performance index in optimal control of fermentation process [103, 110, 206]. Then, we formulate a free time optimal control problem involving the proposed multistage time-delay system and subject to continuous state constraints and control constraint to optimize the fermentation process. Incidentally, optimal control of time-delay systems with fixed terminal time has attracted the attention of many researchers [88, 119, 266, 268]. In contrast, optimal control problems with free terminal time are more difficult than those with fixed terminal time because they require an initial estimation of the unknown terminal time [200]. For this type of optimal control problems involving dynamical systems without time delays, many interesting theoretical results can be found in [168, 174, 222]. For numerical computation, several successful families of algorithms have been developed; see, for example, [47, 140, 141]. Nevertheless, optimal control problems of multistage time-delay systems with free terminal time are rarely considered.

In this chapter, by using a time-scaling transformation, we equivalently transcribe the constrained optimal control problem with free terminal time into the one with fixed terminal time. Furthermore, the transformed optimal control problem

 
C. Liu, Z. Gong, *Optimal Control of Switched Systems Arising in Fermentation Processes*, Springer Optimization and Its Applications 97,
DOI 10.1007/978-3-662-43793-3_10

is approximated by a sequence of parameter optimization problems using the control parameterization method. In addition, the constraint transcription technique is applied to approximate the continuous state constraints by constraints in canonical form. The convergence of this approximation is also established. An improved differential evolution (DE) algorithm is then developed to solve the resultant parameter optimization problems. Numerical results show that the mass of 1,3-PD per unit time is increased considerably and the duration of fermentation is shorted greatly compared with previous results.

The main reference in this chapter is [148].

## 10.2 Controlled Multistage Time-Delay Systems

Under Assumptions 5.2, 5.3, and 9.1, the following controlled multistage time-delay system can be used to describe the fed-batch process

$$\begin{cases} \dot{\boldsymbol{x}}(t) = \boldsymbol{f}^i(\boldsymbol{x}(t), \boldsymbol{x}(t-\alpha), u(t)), \\ u(t) \in U_i,\ t \in (t_{i-1}, t_i],\ i = 1, 2, \ldots, 2N+1, \\ \boldsymbol{x}(0) = \boldsymbol{x}_0, \\ \boldsymbol{x}(t) = \boldsymbol{\phi}(t), t \in [-\alpha,\ 0], \end{cases} \tag{10.1}$$

where $\boldsymbol{x}(t) := (x_1(t), x_2(t), x_3(t), x_4(t), x_5(t), x_6(t))^\top$ is the state vector whose components are, respectively, the extracellular concentrations of biomass, glycerol, 1,3-PD, acetate, and ethanol and the volume of culture fluid at $t$ in the fermenter; $\boldsymbol{x}(t-\alpha)$ is the delayed state; $\alpha$ is a time delay; $u(t)$ is the control function denoting the flow rate of the glycerol; and $t_i, i \in \Lambda := \{1, 2, \ldots, 2N+1\}$, is the switching instant such that $0 = t_0, t_{i-1} < t_i, i \in \Lambda$, and $t_{2N+1} = T$, which is decided a priori in the experiment. In particular, $t_{2j+1}$ is the moment of adding glycerol, at which the fermentation process switches to a feeding process, and $t_{2j+2}$ denotes the moment of ending the flow of glycerol, at which the fermentation process switches to a batch process, $j \in \Lambda_1 := \{0, 1, 2, \ldots, N-1\}$. Moreover, $T$ is the terminal time of the fermentation and is a variable in this chapter, $\boldsymbol{x}_0$ is a given initial state, and $\boldsymbol{\phi}(t) \in C^1([-\alpha, 0], \mathbb{R}^6)$ is a given initial function. Here, $C^1([-\alpha, 0], \mathbb{R}^6)$ is the Banach space of continuously differentiable functions mapping the interval $[-\alpha, 0]$ into $\mathbb{R}^6$. Furthermore, for $t \in (t_{2j}, t_{2j+1}]$, $j \in \Lambda_2 := \{0, 1, \ldots, N\}$,

$$f_\ell^{2j+1}(\boldsymbol{x}(t), \boldsymbol{x}(t-\alpha), u(t)) = q_\ell(\boldsymbol{x}(t))x_1(t-\alpha), \quad \ell = 1, 3, 4, 5, \tag{10.2}$$

$$f_2^{2j+1}(\boldsymbol{x}(t), \boldsymbol{x}(t-\alpha), u(t)) = -q_2(\boldsymbol{x}(t))x_1(t-\alpha), \tag{10.3}$$

and

$$f_6^{2j+1}(\boldsymbol{x}(t), \boldsymbol{x}(t-\alpha), u(t)) = 0; \tag{10.4}$$

for $t \in \left(t_{2j+1}, t_{2j+2}\right], j \in \Lambda_1$,

$$\begin{aligned} & f_\ell^{2j+2}(\boldsymbol{x}(t), \boldsymbol{x}(t-\alpha), u(t)) \\ & \quad = q_\ell(\boldsymbol{x}(t)) x_1(t-\alpha) - D(\boldsymbol{x}(t), u(t)) x_\ell(t), \ \ell = 1,3,4,5, \end{aligned} \tag{10.5}$$

$$\begin{aligned} & f_2^{2j+2}(\boldsymbol{x}(t), \boldsymbol{x}(t-\alpha), u(t)) \\ & \quad = D(\boldsymbol{x}(t), u(t)) \left(\frac{c_{s0}}{1+r} - x_2(t)\right) - q_2(\boldsymbol{x}(t)) x_1(t-\alpha), \end{aligned} \tag{10.6}$$

and

$$f_6^{2j+2}(\boldsymbol{x}(t), \boldsymbol{x}(t-\alpha), u(t)) = (1+r)u(t) \tag{10.7}$$

In (10.5)–(10.7), $c_{s0} > 0$ denotes the concentration of initial feed of glycerol in the medium, and $r > 0$ is the velocity ratio of adding alkali to glycerol. The dilution rate $D(\boldsymbol{x}(t), u(t))$ is defined by

$$D(\boldsymbol{x}(t), u(t)) = \frac{(1+r)u(t)}{x_6(t)}. \tag{10.8}$$

The specific growth rate of cells $q_1(\boldsymbol{x}(t))$, the specific consumption rate of substrate $q_2(\boldsymbol{x}(t))$, and the specific formation rates of products $q_\ell(\boldsymbol{x}(t))$, $\ell = 3,4,5$, are expressed as the following equations:

$$q_1(\boldsymbol{x}(t)) = \frac{\Delta_1 x_2(t)}{x_2(t)+k_1} \prod_{\ell=2}^{5} \left(1 - \frac{x_\ell(t)}{x_\ell^*}\right)^{n_\ell}, \tag{10.9}$$

$$q_2(\boldsymbol{x}(t)) = m_2 + q_1(\boldsymbol{x}(t)) Y_2 + \frac{\Delta_2 x_2(t)}{x_2(t)+k_2}, \tag{10.10}$$

$$q_\ell(\boldsymbol{x}(t)) = -m_\ell + q_1(\boldsymbol{x}(t)) Y_\ell + \frac{\Delta_\ell x_2(t)}{x_2(t)+k_\ell}, \quad \ell = 3,4, \tag{10.11}$$

$$q_5(\boldsymbol{x}(t)) = q_2(\boldsymbol{x}(t)) \left(\frac{c_1}{c_2 + q_1(\boldsymbol{x}(t)) x_2(t)} + \frac{c_3}{c_4 + q_1(\boldsymbol{x}(t)) x_2(t)}\right). \tag{10.12}$$

Under anaerobic conditions, the critical concentrations for the cell growth and the values of parameters in (10.9)–(10.12) are as given in Chap. 9.

Now, define

$$U_i = \begin{cases} [a_i, b_i], & \text{if } i \text{ is even}, \\ \{0\}, & \text{if } i \text{ is odd}, \end{cases} \tag{10.13}$$

where $a_i$ and $b_i$ are positive constants denoting the minimal and the maximal rates of adding glycerol, respectively. Let

$$U := \bigcup_{i=1}^{2N+1} U_i,$$

and assume the terminal time of the fermentation $T$ is also bounded in $[T_{\min}, T_{\max}]$. Thus, we define the class of admissible control functions as

$$\mathcal{U} := \{u \in L_\infty([0, T_{\max}], \mathbb{R}) \mid u(t) \in U_i,\ t \in (t_{i-1}, t_i],\ i = 1, 2, \ldots, 2N+1\}, \tag{10.14}$$

where $L_\infty([0, T_{\max}], \mathbb{R})$ is the Banach space of all essentially bounded functions from $[0, T_{\max}]$ into $\mathbb{R}$.

There exist critical concentrations of biomass, glycerol, 1,3-PD, acetate, and ethanol, outside which cells cease to grow. Hence, it is biologically meaningful to restrict the concentrations of biomass, glycerol, and products within a set $\tilde{W}$ defined as

$$\boldsymbol{x}(t) \in \tilde{W} := \prod_{\ell=1}^{6} \left[x_{*\ell}, x_\ell^*\right], \quad \forall t \in [0, T], \tag{10.15}$$

where $x_{*\ell}, x_\ell^*$, $\ell = 1, 2, \ldots, 5$, are as given in Table 5.1, $x_{*6} = 4$ and $x_6^* = 7$.

For the system (10.1), some important properties are given in the following theorems.

**Theorem 10.1.** *The functions* $\boldsymbol{f}^i : \mathbb{R}_+^6 \times \mathbb{R}_+^6 \times U \to \mathbb{R}^6$, $i = 1, 2, \ldots, 2N+1$, *defined in* (10.2)–(10.7) *satisfy the following conditions:*

(a) $\boldsymbol{f}^i$*, together with its partial derivatives with respect to* $\boldsymbol{x}$*,* $\boldsymbol{y}$*, and* $u$*, are continuous on* $\mathbb{R}_+^6 \times \mathbb{R}_+^6 \times U$*.*

(b) *There exists a constant* $K > 0$ *such that*

$$\|\boldsymbol{f}^i(\boldsymbol{x}, \boldsymbol{y}, u)\| \leqslant K(1 + \|\boldsymbol{x}\| + \|\boldsymbol{y}\|), \forall (\boldsymbol{x}, \boldsymbol{y}, u) \in \mathbb{R}_+^6 \times \mathbb{R}_+^6 \times U_i, \tag{10.16}$$

*where* $\|\cdot\|$ *denotes the Euclidean norm.*

*Proof.* (a) This conclusion can be obtained by the expression of $\boldsymbol{f}^i$ in (10.2)–(10.7).

(b) The result can be proved in a similar manner to the proof that is given for Proposition 5.1 in Chap. 5. □

**Theorem 10.2.** *For each* $u \in \mathcal{U}$ *and* $T \in [T_{\min}, T_{\max}]$*, the system* (10.1) *has a unique continuous solution, denoted by* $\boldsymbol{x}(\cdot|u, T)$*, on* $[-\alpha, T]$*. Furthermore,* $\boldsymbol{x}(\cdot|u, T)$ *satisfies that*

$$x(t|u,T) = x(t_{i-1}|u,T) + \int_{t_{i-1}}^{t} f^i(x(s|u,T), x(s-\alpha|u,T), u(s))\mathrm{d}s,$$
$$\forall t \in (t_{i-1}, t_i],\ i \in \Lambda. \tag{10.17}$$

*and* $x(t|u,T) = \phi(t),\ \forall t \in [-\alpha, 0]$.

*Proof.* The proof can be obtained by Theorem 10.1 and the theory of delay-differential equations [95]. □

**Theorem 10.3.** *Given the initial function* $\phi(t) \in C^1\left([-\alpha,0], \mathbb{R}^6_+\right)$ *and the initial state* $x_0$*, the unique solution* $x(\cdot|u,T)$ *of the system* (10.1) *is uniformly bounded.*

*Proof.* For each $u \in \mathcal{U}$ and $T \in [T_{\min}, T_{\max}]$, since $\phi(t)$ is continuous on $[-\alpha, 0]$, there exists a constant $M' \geqslant 0$ such that

$$\sup\{\|\phi(t)\| \mid t \in [-\alpha, 0]\} \leqslant M'.$$

Thus,

$$\|x(t|u,T)\| \leqslant M',\ \forall t \in [-\alpha, 0].$$

In view of Theorems 10.1 and 10.2, we obtain that

$$\begin{aligned}\|x(t|u,T)\| &\leqslant \|x_0\| + \sum_{j=1}^{i-1}\int_{t_{j-1}}^{t_j} \|f^j(x(s|u,T), x(s-\alpha|u,T), u(s))\|\mathrm{d}s \\ &\quad + \int_{t_{i-1}}^{t} \|f^i(x(s|u,T), x(s-\alpha|u,T), u(s))\|\mathrm{d}s \\ &\leqslant \|x_0\| + \int_0^t K(1 + \|x(s|u,T)\| + \|x(s-\alpha|u,T)\|)\mathrm{d}s \\ &\leqslant M' + K\alpha M' + K\int_0^t (1 + 2\|x(s|u,T)\|)\mathrm{d}s, \quad \forall t \in (0, T].\end{aligned}$$

By Lemma 4.1, it follows that

$$\|x(t|u,T)\| \leqslant \left(M' + K\alpha M' + KT_{\max}\right)\exp(2KT_{\max}), \quad \forall t \in (0, T].$$

Therefore,

$$\|x(t|u,T)\| \leqslant M,\ \forall t \in [-\alpha, T],$$

where $M := \max\{M', (M' + K\alpha M' + KT_{\max})\exp(2KT_{\max})\}$. □

## 10.3 Constrained Optimal Control Problems

In fed-batch process, it is desired that the value of the target product 1,3-PD should be maximized at the end of the process and, at the same time, the operation costs should be reduced. In particular, both the flow rate of glycerol and the terminal time of the fermentation play key roles in achieving the objective. Thus, we take the mass of 1,3-PD per unit time in the fed-batch process as the cost functional which can be formulated as

$$J(u,T) = \frac{x_3(T|u,T)x_6(T|u,T)}{T}, \tag{10.18}$$

where $x_3(T|u,T)$ and $x_6(T|u,T)$ are, respectively, the third and the sixth components of the solution to the system (10.1) at terminal time $T$.

Now, we can formally state the optimal control problem as

**Problem 10.1.** Given the system (10.1), find a control $u \in \mathcal{U}$ and a terminal time $T \in [T_{\min}, T_{\max}]$ such that the state constraint (10.15) is satisfied and the cost functional (10.18) is maximized.

Note that Problem 10.1 is of nonstandard feature because it has not fixed terminal time but free terminal time. It is difficult to solve Problem 10.1 using existing numerical techniques [143, 150, 151]. The main difficulty is the implicit dependence of the system state on the terminal time. We now employ a time-scaling transformation from $t \in [0,T]$ to $s \in [0,1]$ as follows:

$$t = Ts. \tag{10.19}$$

Then, let $\tilde{\boldsymbol{x}}(s) := \boldsymbol{x}(t(s))$, $\tilde{u}(s) := u(t(s))$, $\tilde{\alpha} := \frac{\alpha}{T}$, $\boldsymbol{h}^i(\tilde{\boldsymbol{x}}(s), \tilde{\boldsymbol{x}}(s-\tilde{\alpha}), \tilde{u}(s), T) := T\boldsymbol{f}^i(\tilde{\boldsymbol{x}}(s), \tilde{\boldsymbol{x}}(s-\tilde{\alpha}), \tilde{u}(s))$ and $\tilde{\boldsymbol{\phi}}(s) := \boldsymbol{\phi}(t(s))$. As a result, the system (10.1) takes the form

$$\begin{cases} \dot{\tilde{\boldsymbol{x}}}(s) = \boldsymbol{h}^i(\tilde{\boldsymbol{x}}(s), \tilde{\boldsymbol{x}}(s-\tilde{\alpha}), \tilde{u}(s), T), \\ \tilde{u}(s) \in U_i, \ s \in (s_{i-1}, s_i], \ i = 1, 2, \ldots, 2N+1, \\ \tilde{\boldsymbol{x}}(0) = \boldsymbol{x}_0, \\ \tilde{\boldsymbol{x}}(s) = \tilde{\boldsymbol{\phi}}(s), s \in [-\tilde{\alpha}, 0]. \end{cases} \tag{10.20}$$

Furthermore, the switching instants $t_i$ in the original time are transcribed into $s_i = \frac{t_i}{T}$, $i = 1, 2, \ldots, 2N$. Now, let $\tilde{\boldsymbol{x}}(\cdot|\tilde{u}, T)$ be the solution of the transformed system (10.20). Accordingly, the class of admissible control functions can be transcribed into $\tilde{\mathcal{U}}$ and the state constraint (10.15) can be rewritten as

$$\tilde{\boldsymbol{x}}(s|\tilde{u}, T) \in \tilde{W}. \tag{10.21}$$

Therefore, Problem 10.1 can be transcribed to the following equivalent problem with fixed terminal time.

**Problem 10.2.** Subject to the system (10.20), find a control $\tilde{u} \in \tilde{\mathcal{U}}$ and a terminal time $T \in [T_{\min}, T_{\max}]$ such that the state constraint (10.21) is satisfied and the cost functional

$$\tilde{J}(u, T) = \frac{\tilde{x}_3(1|\tilde{u}, T)\tilde{x}_6(1|\tilde{u}, T)}{T} \tag{10.22}$$

is maximized.

By the similar arguments as those given for Theorem 5.3 in Chap. 5, we confirm the existence of the optimal solution for Problem 10.2.

**Theorem 10.4.** *Problem* 10.2 *has at least one optimal solution.*

## 10.4 Computational Approaches

Problem 10.2 is essentially a constrained optimal control problem. It is known that computational schemes based on the control parameterization technique are normally very efficient in solving optimal control problems [244, 257, 265, 266]. In this section, we will develop a computational method using the control parameterization method in conjunction with an improved DE algorithm to solve Problem 10.2.

For each $p_i \geqslant 1, i \in \{1, 2, \ldots, 2N+1\}$, let the subinterval $[s_{i-1}, s_i]$ be partition into $n_{p_i}$ subintervals with $n_{p_i} + 1$ partition points such that

$$s_{i-1} = p_0^i \leqslant p_1^i, \cdots, \leqslant p_{n_{p_i}}^i = s_i,$$

where $n_{p_i}$ is chosen such that $n_{p_i+1} \geqslant n_{p_i}$. Then, the control can be approximated as

$$\tilde{u}^p(s) = \sum_{i=1}^{2N+1} \sum_{k=1}^{n_{p_i}} \sigma^{i,k} \chi_{(p_{k-1}^i, p_k^i]}(s), \tag{10.23}$$

where $\chi_{(p_{k-1}^i, p_k^i]}$ is the indicator function on the interval $\left(p_{k-1}^i, p_k^i\right]$ defined by

$$\chi_{(p_{k-1}^i, p_k^i]}(s) = \begin{cases} 1, & s \in \left(p_{k-1}^i, p_k^i\right], \\ 0, & \text{otherwise.} \end{cases}$$

Let $\boldsymbol{\sigma}^p := \left((\sigma^1)^\top, (\sigma^2)^\top, \ldots, (\boldsymbol{\sigma}^{2N+1})^\top\right)^\top \in \mathbb{R}^l$, where $\boldsymbol{\sigma}^i := (\sigma^{i,1}, \sigma^{i,2}, \ldots, \sigma^{i,n_{p_i}})^\top$ defines the heights of the approximate control (10.23) and $l := \sum_{i=1}^{2N+1} n_{p_i}$. From (10.14), it is clear that

$$\sigma^{i,k} \in U_i,\ k = 1, 2, \ldots, n_{p_i};\ i = 1, 2, \ldots, 2N + 1. \tag{10.24}$$

Let $\Xi^p$ be the set of all those $\boldsymbol{\sigma}^p$ satisfying the constraint (10.24). Furthermore, denote the solution of the system (10.20) replacing the control function $\tilde{u}$ with $\tilde{u}^p$ by $\bar{\boldsymbol{x}}(\cdot|\boldsymbol{\sigma}^p, T)$. Accordingly, the state constraint (10.21) becomes

$$\bar{\boldsymbol{x}}(s|\boldsymbol{\sigma}^p, T) \in \tilde{W}. \tag{10.25}$$

Thus, we may specify the approximate problem as follows.

**Problem 10.3.** Given the replaced system of system (10.20), find a control parameter vector $\boldsymbol{\sigma}^p \in \Xi^p$ and a terminal time $T \in [T_{\min}, T_{\max}]$ such that the state constraint (10.25) is satisfied and the cost functional

$$\bar{J}(\boldsymbol{\sigma}^p, T) = \frac{\bar{x}_3(1|\boldsymbol{\sigma}^p, T)\bar{x}_6(1|\boldsymbol{\sigma}^p, T)}{T} \tag{10.26}$$

is maximized.

Note that Problem 10.2 can be approximated by a sequence of parameter optimization problems with continuous state inequality constraint (10.25). However, it is difficult to deal with the continuous state inequality constraint in numerically solving the optimization problem. For this reason, let

$$g_\ell(\bar{\boldsymbol{x}}(s|\boldsymbol{\sigma}^p, T)) = x_\ell^* - \bar{x}_\ell(s|\boldsymbol{\sigma}^p, T),$$
$$g_{6+\ell}(\bar{\boldsymbol{x}}(s|\boldsymbol{\sigma}^p, T)) = \bar{x}_\ell(s|\boldsymbol{\sigma}^p, T) - x_{*\ell},\ \ell = 1, 2, \ldots, 6.$$

Then, the state constraint (10.25) is equivalently transcribed into

$$G(\boldsymbol{\sigma}^p, T) = 0, \tag{10.27}$$

where $G(\boldsymbol{\sigma}^p, T) := \sum_{l=1}^{12} \int_0^1 \min\{0, g_l(\bar{\boldsymbol{x}}(s|\boldsymbol{\sigma}^p, T))\} \mathrm{d}s$. However, the equality constraint (10.27) is non-differentiable at the points when $g_l = 0, l \in \{1, 2, \ldots, 12\}$. Using the method given in Chap. 6, we approximate the state constraint (10.25) as the following inequality constraint:

$$\bar{G}_{\varepsilon,\delta}(\boldsymbol{\sigma}^p, T) := \delta + \sum_{l=1}^{12} \int_0^1 \varphi_\varepsilon(g_l(\bar{\boldsymbol{x}}(s|\boldsymbol{\sigma}^p, T))) \mathrm{d}s \geq 0, \tag{10.28}$$

where $\varepsilon > 0$, $\delta > 0$ and

$$\varphi_\varepsilon(\eta) = \begin{cases} \eta, & \text{if } \eta < -\varepsilon, \\ -\dfrac{(\eta-\varepsilon)^2}{4\varepsilon}, & \text{if } -\varepsilon \leqslant \eta \leqslant \varepsilon, \\ 0, & \text{if } \eta > \varepsilon. \end{cases} \tag{10.29}$$

It should be noted that this function is obtained by smoothing out the sharp corner of the function $\min\{0, g_l(\bar{\boldsymbol{x}}(s|\boldsymbol{\sigma}^p, T))\}$. Consequently, Problem 10.3 is approximated by a sequence of Problems $\{(\mathrm{EP}_{\varepsilon,\delta}(\mathrm{p}))\}$ defined by replacing the state constraint (10.25) with the inequality constraint (10.28). Under appropriate assumptions, it shown in Lemma 8.3.3 of [240] that for all $\varepsilon > 0$, there exists a $\delta(\varepsilon) > 0$ such that for all $\delta, 0 < \delta < \delta(\varepsilon)$, if an admissible pair $(\boldsymbol{\sigma}^p, T) \in \Xi^p \times [T_{\min}, T_{\max}]$ satisfies the inequality constraint (10.28), then it also satisfies the state constraint (10.25).

In the numerical computation, the gradients of $\bar{G}_{\varepsilon,\delta}(\boldsymbol{\sigma}^p, T)$ with respect to $\boldsymbol{\sigma}^p$ and $T$ are required. We will develop a scheme using the auxiliary systems, which is similar to the method given for Theorem 9.9, for computing the gradients of the constraint $\bar{G}_{\varepsilon,\delta}(\cdot,\cdot)$ in the following theorems.

**Theorem 10.5.** *For each $\varepsilon > 0$ and $\delta > 0$, the gradients of the constraint $\bar{G}_{\varepsilon,\delta}(\boldsymbol{\sigma}^p, T)$ defined in (10.28) with respect to $\boldsymbol{\sigma}^p$ are*

$$\frac{\partial \bar{G}_{\varepsilon,\delta}(\boldsymbol{\sigma}^p, T)}{\partial \sigma^{i,k}} = \sum_{l=1}^{12} \int_0^1 \frac{\partial \varphi_\varepsilon(g_l(\bar{\boldsymbol{x}}(s|\boldsymbol{\sigma}^p, T)))}{\partial g_l} \frac{\partial g_l(\bar{\boldsymbol{x}}(s|\boldsymbol{\sigma}^p, T))}{\partial \bar{\boldsymbol{x}}} \boldsymbol{\xi}^{i,k}(s)\mathrm{d}s$$
$$k = 1, 2 \ldots, n_{p_i};\ i = 1, 2, \ldots, 2N+1, \tag{10.30}$$

*where $\boldsymbol{\xi}^{i,k}(s)$ are the solutions of the following time-delay systems:*

$$\dot{\boldsymbol{\xi}}^{i,k}(s) = (1-\upsilon_i)(1-\varsigma_{i,k}(s))\Bigg\{ \frac{\partial \boldsymbol{h}^i(\bar{\boldsymbol{x}}(s|\boldsymbol{\sigma}^p, T), \bar{\boldsymbol{x}}(s-\tilde{\alpha}|\boldsymbol{\sigma}^p, T), \boldsymbol{\sigma}^p, T)}{\partial \bar{\boldsymbol{x}}(s)} \boldsymbol{\xi}^{i,k}(s)$$
$$+ \frac{\partial \boldsymbol{h}^i(\bar{\boldsymbol{x}}(s|\boldsymbol{\sigma}^p, T), \bar{\boldsymbol{x}}(s-\tilde{\alpha}|\boldsymbol{\sigma}^p, T), \boldsymbol{\sigma}^p, T)}{\partial \bar{\boldsymbol{x}}(s-\tilde{\alpha})} \boldsymbol{\xi}^{i,k}(s-\tilde{\alpha}) +$$
$$\frac{\partial \boldsymbol{h}^i(\bar{\boldsymbol{x}}(s|\boldsymbol{\sigma}^p, T), \bar{\boldsymbol{x}}(s-\tilde{\alpha}|\boldsymbol{\sigma}^p, T), \boldsymbol{\sigma}^p, T)}{\partial \sigma^{i,k}} \Bigg\},\ \forall s \in (s_{i-1}, s_i], \tag{10.31}$$

*with*

$$\boldsymbol{\xi}^{i,k}(s) = \mathbf{0},\ s \in [-\tilde{\alpha}, 0], \tag{10.32}$$

$$\upsilon_i = \begin{cases} 1, & i \text{ is odd}, \\ 0, & \text{otherwise}, \end{cases} \tag{10.33}$$

*and*

$$\varsigma_{i,k}(s) = \begin{cases} 1, & s \leqslant s_{k-1}^i, \\ 0, & \textit{otherwise}. \end{cases} \tag{10.34}$$

*Proof.* For each $\epsilon \geqslant 0$, define

$$\boldsymbol{\sigma}^{p,\epsilon} = \left((\boldsymbol{\sigma}^1)^\top, \ldots, (\sigma^{i,1}, \ldots, \sigma^{i,k} + \epsilon, \ldots, \sigma^{i,n_{p_i}}), \ldots, (\boldsymbol{\sigma}^{2N+1})^\top\right)^\top \in \Xi^p.$$

For brevity, let $\bar{\boldsymbol{x}}(s)$ and $\bar{\boldsymbol{x}}^\epsilon(s)$, $\forall s \in (s_{i-1}, s_i]$, denote the solutions of the system (10.20) with $u^p$ corresponding to $\boldsymbol{\sigma}^p$ and $\boldsymbol{\sigma}^p(\epsilon)$, respectively. Clearly, we have

$$\begin{aligned}\bar{\boldsymbol{x}}(s) = \boldsymbol{x}_0 &+ \sum_{j=1}^{i-1} \int_{s_{j-1}}^{s_j} \boldsymbol{h}^j(\bar{\boldsymbol{x}}(\vartheta), \bar{\boldsymbol{x}}(\vartheta - \tilde{\alpha}), \boldsymbol{\sigma}^p, T)\mathrm{d}\vartheta \\ &+ \int_{s_{i-1}}^{s} \boldsymbol{h}^i(\bar{\boldsymbol{x}}(\vartheta), \bar{\boldsymbol{x}}(\vartheta - \tilde{\alpha}), \boldsymbol{\sigma}^p, T)\mathrm{d}\vartheta\end{aligned}$$

and

$$\begin{aligned}\bar{\boldsymbol{x}}^\epsilon(s) = \boldsymbol{x}_0 &+ \sum_{j=1}^{i-1} \int_{s_{j-1}}^{s_j} \boldsymbol{h}^j(\bar{\boldsymbol{x}}^\epsilon(\vartheta), \bar{\boldsymbol{x}}^\epsilon(\vartheta - \tilde{\alpha}), \boldsymbol{\sigma}^{p,\epsilon}, T)\mathrm{d}\vartheta \\ &+ \int_{s_{i-1}}^{s} \boldsymbol{h}^i(\bar{\boldsymbol{x}}^\epsilon(\vartheta), \bar{\boldsymbol{x}}^\epsilon(\vartheta - \tilde{\alpha}), \boldsymbol{\sigma}^{p,\epsilon}, T)\mathrm{d}\vartheta.\end{aligned}$$

Consequently, if $i$ is odd or $s \leqslant s_{k-1}^i$, then

$$\frac{\partial \bar{\boldsymbol{x}}(s)}{\partial \sigma^{i,k}} = \boldsymbol{0}. \tag{10.35}$$

Otherwise,

$$\begin{aligned}\frac{\partial \bar{\boldsymbol{x}}(s)}{\partial \sigma^{i,k}} = \int_{s_{k-1}^i}^{s} \Bigg\{ &\frac{\partial \boldsymbol{h}^i(\bar{\boldsymbol{x}}(\vartheta), \bar{\boldsymbol{x}}(\vartheta - \tilde{\alpha}), \boldsymbol{\sigma}^p, T)}{\partial \bar{\boldsymbol{x}}(\vartheta)} \frac{\partial \bar{\boldsymbol{x}}(\vartheta)}{\partial \sigma^{i,k}} + \frac{\partial \boldsymbol{h}^i(\bar{\boldsymbol{x}}(\vartheta), \bar{\boldsymbol{x}}(\vartheta - \tilde{\alpha}), \boldsymbol{\sigma}^p, T)}{\partial \bar{\boldsymbol{x}}(\vartheta - \tilde{\alpha})} \\ &\times \frac{\partial \bar{\boldsymbol{x}}(\vartheta - \tilde{\alpha})}{\partial \sigma^{i,k}} + \frac{\partial \boldsymbol{h}^i(\bar{\boldsymbol{x}}(\vartheta), \bar{\boldsymbol{x}}(\vartheta - \tilde{\alpha}), \boldsymbol{\sigma}^p, T)}{\partial \sigma^{i,k}} \Bigg\} \mathrm{d}\vartheta.\end{aligned} \tag{10.36}$$

Differentiating (10.35) and (10.36) with respect to time yields

$$\frac{\mathrm{d}}{\mathrm{d}s}\left\{\frac{\partial \bar{\boldsymbol{x}}(s)}{\partial \sigma^{i,k}}\right\} = (1-\upsilon_i)(1-\varsigma_{i,k}(s))\left\{\frac{\partial \boldsymbol{h}^i(\bar{\boldsymbol{x}}(s), \bar{\boldsymbol{x}}(s-\tilde{\alpha}), \boldsymbol{\sigma}^p, T)}{\partial \bar{\boldsymbol{x}}(s)}\frac{\partial \bar{\boldsymbol{x}}(s)}{\partial \sigma^{i,k}} + \right.$$
$$\left.\frac{\partial \boldsymbol{h}^i(\bar{\boldsymbol{x}}(s), \bar{\boldsymbol{x}}(s-\tilde{\alpha}), \boldsymbol{\sigma}^p, T)}{\partial \bar{\boldsymbol{x}}(s-\tilde{\alpha})}\frac{\partial \bar{\boldsymbol{x}}(s-\tilde{\alpha})}{\partial \sigma^{i,k}} + \frac{\partial \boldsymbol{h}^i(\bar{\boldsymbol{x}}(s), \bar{\boldsymbol{x}}(s-\tilde{\alpha}), \boldsymbol{\sigma}^p, T)}{\partial \sigma^{i,k}}\right\},$$
$$\forall s \in (s_{i-1}, s_i],\ i = 1, 2, \ldots, 2N+1,$$

where $\upsilon_i$ and $\varsigma_{i,k}(s)$ are as defined in (10.33) and (10.34), respectively. Furthermore,

$$\frac{\partial \bar{\boldsymbol{x}}(0)}{\partial \sigma^{i,k}} = \frac{\partial}{\partial \sigma^{i,k}}\left\{\boldsymbol{x}_0\right\} = \mathbf{0},$$
$$\frac{\partial \bar{\boldsymbol{x}}(s)}{\partial \sigma^{i,k}} = \frac{\partial \tilde{\boldsymbol{\phi}}(s)}{\partial \sigma^{i,k}} = \mathbf{0},\ s \in [-\tilde{\alpha}, 0].$$

Hence, defining

$$\boldsymbol{\xi}^{i,k}(s) = \frac{\partial \bar{\boldsymbol{x}}(s)}{\partial \sigma^{i,k}}$$

and differentiating $\bar{G}_{\varepsilon,\delta}(\boldsymbol{\sigma}^p, T)$ with respect to $\boldsymbol{\sigma}^p$, we obtain the conclusion (10.30). □

Define

$$\bar{\boldsymbol{\chi}}(s) := \begin{cases} \dot{\tilde{\boldsymbol{\phi}}}(s), & \text{if } s \leqslant 0, \\ \boldsymbol{h}^i(\bar{\boldsymbol{x}}(s), \bar{\boldsymbol{x}}(s-\tilde{\alpha}), \boldsymbol{\sigma}^p, T), & \text{if } s \in (s_{i-1}, s_i] \text{ for some } i \in \Lambda. \end{cases}$$

**Theorem 10.6.** *For each $\varepsilon > 0$ and $\delta > 0$, the gradient of the constraint $\bar{G}_{\varepsilon,\delta}(\boldsymbol{\sigma}^p, T)$ defined in* (10.28) *with respect to $T$ is*

$$\frac{\partial \bar{G}_{\varepsilon,\delta}(\boldsymbol{\sigma}^p, T)}{\partial T} = \sum_{l=1}^{12}\int_0^1 \frac{\partial \varphi_\varepsilon(g_l(\bar{\boldsymbol{x}}(s|\boldsymbol{\sigma}^p, T)))}{\partial g_l}\frac{\partial g_l(\bar{\boldsymbol{x}}(s|\boldsymbol{\sigma}^p, T))}{\partial \bar{\boldsymbol{x}}}\boldsymbol{\zeta}(s)\mathrm{d}s$$

*where $\boldsymbol{\zeta}(s)$ is the solution of the following time-delay system:*

$$\dot{\boldsymbol{\zeta}}(s) = \frac{\partial \boldsymbol{h}^i(\bar{\boldsymbol{x}}(s|\boldsymbol{\sigma}^p, T), \bar{\boldsymbol{x}}(s-\tilde{\alpha}|\boldsymbol{\sigma}^p, T), \boldsymbol{\sigma}^p, T)}{\partial \bar{\boldsymbol{x}}(s)}\boldsymbol{\zeta}(s)$$
$$+ \frac{\partial \boldsymbol{h}^i(\bar{\boldsymbol{x}}(s|\boldsymbol{\sigma}^p, T), \bar{\boldsymbol{x}}(s-\tilde{\alpha}|\boldsymbol{\sigma}^p, T), \boldsymbol{\sigma}^p, T)}{\partial \bar{\boldsymbol{x}}(s-\tilde{\alpha})}\boldsymbol{\zeta}(s-\tilde{\alpha})$$
$$+ \frac{\tilde{\alpha}}{T}\frac{\partial \boldsymbol{h}^i(\bar{\boldsymbol{x}}(s|\boldsymbol{\sigma}^p, T), \bar{\boldsymbol{x}}(s-\tilde{\alpha}|\boldsymbol{\sigma}^p, T), \boldsymbol{\sigma}^p, T)}{\partial \bar{\boldsymbol{x}}(s-\tilde{\alpha})}\bar{\boldsymbol{\chi}}(s-\tilde{\alpha})$$
$$+ \boldsymbol{f}^i(\bar{\boldsymbol{x}}(s|\boldsymbol{\sigma}^p, T), \bar{\boldsymbol{x}}(s-\tilde{\alpha}|\boldsymbol{\sigma}^p, T), \boldsymbol{\sigma}^p),$$
$$\forall s \in (s_{i-1}, s_i],\ i = 1, 2, \ldots, 2N+1,$$

*with*

$$\boldsymbol{\zeta}(s) = \mathbf{0}, \; s \in [-\tilde{\alpha}, 0].$$

*Proof.* The proof can be completed using a similar method given for Theorem 10.5. □

Based on the above theorems, Problem 10.2 can be solved by a sequence of approximation problems $\{(\mathrm{EP}_{\varepsilon,\delta}(\mathrm{p}))\}$. Each of $\{(\mathrm{EP}_{\varepsilon,\delta}(\mathrm{p}))\}$ is a smooth mathematical programming problem which can be solved by gradient-based techniques [38,47,240]. However, the gradient-based techniques are only designed to find local optima. Furthermore, in solving $\{(\mathrm{EP}_{\varepsilon,\delta}(\mathrm{p}))\}$, the evaluation of candidate feeding rate as well as the terminal time is a computationally expensive operation because of solving the system (10.1). As a result, finding the global optimum or a good suboptimal solution with traditional search or optimization techniques based on natural phenomenon such as genetic algorithm [104], simulation annealing [122], and evolution strategies [219] is too consuming or even impossible within the time available.

DE algorithm has been used in the recent past to solve many engineering problems; see, for example, [50, 251]. When using the DE to optimize a function, an acceptable trade-off between convergence and robustness must generally be determined. To increase the convergence without compromising with the robustness, a modified differential evolution (MDE) is developed to solve unconstrained optimization problems encountered in chemical engineering [12]. The basic operations of MDE are similar to those of conventional DE algorithm. However, it can use a smaller population size to achieve a high probability of obtaining the optimum [12]. Nevertheless, the $(\mathrm{EP}_{\varepsilon,\delta}(\mathrm{p}))$ is a nonlinear optimization problem with constraints in state and control parameters, which MDE cannot be applied directly to solve it. Hence, the following strategies are added to the MDE algorithm in [12].

(I) (Handling the control constraints) If there is a bound violation for a parameter in the $\iota$th individual at the $\kappa$th step, then that parameter is generated randomly between the given lower and upper bounds using the following equations:

$$\sigma^p_{\iota,\jmath}(\kappa) = \sigma^p_{\mathrm{low},\jmath} + r^1_{\iota,\jmath} \cdot \left(\sigma^p_{\mathrm{upp},\jmath} - \sigma^p_{\mathrm{low},\jmath}\right), \quad \jmath = 1, \ldots, l,$$

and

$$T_\iota(\kappa) = T_{\min} + r^2_\iota \cdot (T_{\max} - T_{\min}), \quad \iota = 1, 2, \ldots, N_p,$$

where $\boldsymbol{\sigma}^p_{\mathrm{low}}$ and $\boldsymbol{\sigma}^p_{\mathrm{upp}}$ are, respectively, the lower and upper bounds of the control parameter which can be obtained by (10.24) and $r^1_{\iota,\jmath}$ and $r^2_\iota$ are random numbers taken from [0, 1].

(II) (Dealing with the continuous state constraints) For the parameter of the $\iota$th individual at the $\kappa$th step, test the value of $G\left(\boldsymbol{\sigma}^p_\iota(\kappa), T_\iota(\kappa)\right)$. If

$G\left(\sigma_\iota^p(\kappa), T_\iota(\kappa)\right) = 0$, then the parameter is feasible. Otherwise, move the parameter toward the feasible region using the gradient information $\dfrac{\partial \bar{G}_{\varepsilon,\delta}\left(\sigma_\iota^p(\kappa), T_\iota(\kappa)\right)}{\partial \sigma_\iota^p(\kappa)}$ and $\dfrac{\partial \bar{G}_{\varepsilon,\delta}\left(\sigma_\iota^p(\kappa), T_\iota(\kappa)\right)}{\partial T_\iota(\kappa)}$ with Armijo line searches.

(III) (Stopping criteria) The algorithm stops when the maximal iteration $M_p$ is reached.

In view of Theorems 10.5 and 10.6, the following algorithm can now be used to generate an approximate optimal control of Problem 10.1.

**Algorithm 10.1.**

**Step 1.** Choose initial values of $\varepsilon, \delta$ and $(\sigma^p, T)$; set parameters $0 < \beta_1 < 1$, $0 < \beta_2 < 1$, $\bar{\varepsilon}$, and $\bar{\delta}$.

**Step 2.** Solve approximate problem ($EP_{\varepsilon,\delta}$(p)) using the improved DE algorithm to give $(\sigma_{\varepsilon,\delta}^{p,*}, T_{\varepsilon,\delta}^*)$.

**Step 3.** Check the value of $G(\sigma_{\varepsilon,\delta}^{p,*}, T_{\varepsilon,\delta}^*)$.

**Step 4.** If $G(\sigma_{\varepsilon,\delta}^{p,*}, T_{\varepsilon,\delta}^*) = 0$, then go to Step 6. Otherwise, set $\delta := \beta_1\delta$. If $\delta < \bar{\delta}$, then go to Step 5. Otherwise, go to Step 2.

**Step 5.** Set $\varepsilon := \beta_2\varepsilon$. If $\varepsilon \geqslant \bar{\varepsilon}$, then go to Step 3. Otherwise go to Step 6.

**Step 6.** If $\min_{i\in\{1,2,\ldots,2N+1\}} n_{p_i} \geqslant \bar{P}$, where $\bar{P}$ is a predefined positive constant, then go to Step 7. Otherwise, go to Step 2 with $n_{p_i}$ increased to $n_{p_i+1}$ for each $i$.

**Step 7.** Construct $(u^{p,*}, T^*)$ from $(\sigma_{\varepsilon,\delta}^{p,*}, T_{\varepsilon,\delta}^*)$ by (10.19) and (10.23) and stop.

At the conclusion of Steps 1–7, $(u^{p,*}, T^*)$ is an approximate optimal solution of Problem 10.1.

## 10.5 Numerical Results

In numerical simulation, the reactant composition, cultivation conditions, and the determination of biomass, substrate, and metabolites have been reported in [48]. To numerically solve the system (10.1), the initial state, the velocity ratio of adding alkali to glycerol, the concentration of initial feed glycerol, and time delay are $\boldsymbol{x}_0 = (0.1115\,\mathrm{g\,L^{-1}}, 495\,\mathrm{mmol\,L^{-1}}, 0, 0, 0, 5\,\mathrm{L})^\top$, $r = 0.75$, $c_{s0} = 10{,}762\,\mathrm{mmol\,L^{-1}}$, and $\alpha = 0.4652\,\mathrm{h}$, respectively. In addition, the initial function $\boldsymbol{\phi}(t)$ is interpolated by the cubic spline [189] of the experimental data, $T_{\min} = 11\,\mathrm{h}$, and $T_{\max} = 24.16\,\mathrm{h}$.

In computational process, we use the same switching instants and feeding rate settings as those used to obtain the experimental results to optimize the feeding rates and the terminal time. More specifically, the maximal duration of fed-batch process is partitioned into the first batch phase (Bat. Ph.) and phases I–IX (Phs. I–IX) according to the number of switchings. The same feeding strategies are adopted in each one of Ph. I to Ph. IX. Furthermore, $t_1 = 5.33\,\mathrm{h}$, the feeding moment $t_{2j+1}$,

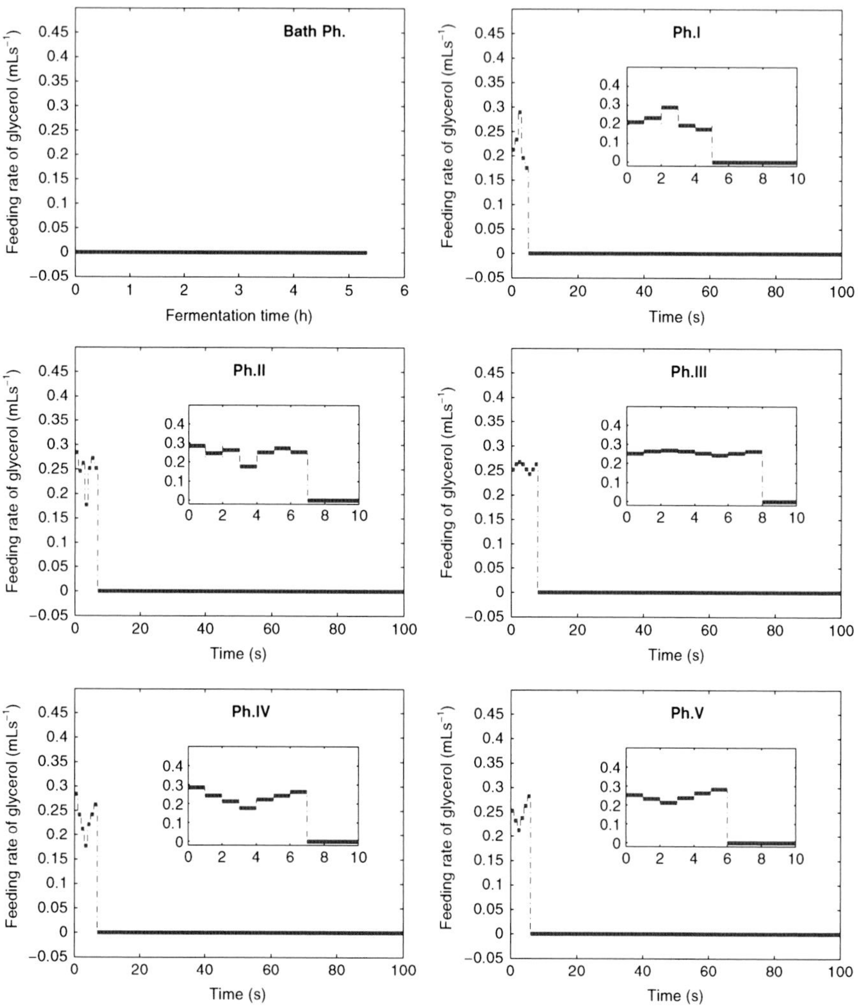


**Fig. 10.1** The optimal feeding strategy of glycerol in fed-batch process

and the end of the feeding moment $t_{2j+2}$ are determined by the experiment. Namely, the durations of the feeding processes in Phs. I–IX are $5, 7, 8, 7, 6, 4, 3, 2$, and 1 s in each 100 s, leaving $95, 93, 92, 93, 94, 96, 97, 98$, and 99 s for batch processes, respectively. This is also done for the computational time consideration since there are a total of 1,355 switchings in the maximal duration of fed-batch process. Moreover, the bounds of feeding rates in Phs. I–IX are as listed in Table 5.2.

In the improved DE algorithm, the size of population $Np$, the maximal iteration $Mp$, the scaling factor $F$, and the crossover constant $CR$ are, respectively, 100,

200, 0.5, and 0.8. In Algorithm 10.1, the initial values of $u$ and $T$ are chosen as those in Chap. 5, in which the corresponding $N = 677$. The other parameters $\varepsilon, \delta, \beta_1, \beta_2, \bar{\varepsilon}, \bar{\delta}$, and $\bar{P}$ are chosen as $0.1, 0.01, 0.1, 0.01, 1.0 \times 10^{-8}, 1.0 \times 10^{-7}$, and 1, respectively.

Applying Algorithm 10.1 to Problem 10.1, we obtain the optimal terminal time $T^* = 13.6694$ h which is much shorter than 19.83 h in the experiment and 21.1078 h in Chap. 5. Moreover, under the optimal terminal time, the corresponding optimal number of switchings is $N^* = 282$. This is very interesting for the biochemical engineer to reduce the operation costs in the fed-batch process. As a result, the optimal feeding rates of glycerol in Bat. Ph. and Phs. I–V are shown in Fig. 10.1. Here, all the computations are performed in Microsoft Visual C++ 6.0 and numerical results are plotted by Matlab 7.10.0. In particular, the combination of the fourth-order Runge–Kutta integration scheme with the cubic spline interpolation [189] is used to integrate the delay-differential equations with the relative error tolerance $10^{-6}$. In detail, the line in the first subfigure of Fig. 10.1 indicates the feeding rate of glycerol, which is identically equal to zero, and the time duration in the Bat. Ph. Accordingly, the lines in the next 5 subfigures illustrate the feeding rates of glycerol in conjunction with time durations of a feeding process and its succeeding batch process in Phs. I–V, respectively. To show the feeding rates of glycerol for Phs. I–V better, 5 small subfigures are also incorporated in the corresponding 5 subfigures, respectively.

Under the obtained optimal feeding rates and the optimal terminal time, the mass of 1,3-PD per unit time is 287.173 mmol $h^{-1}$ which is increased by 7.76 % in comparison with experimental result 266.496 mmol $h^{-1}$ and by 6.04 % compared with the computational result 270.827 mmol $h^{-1}$ in Chap. 5. The optimal computed profile of the mass of 1,3-PD per unit time is depicted by solid curve in Fig. 10.2. In addition, the computational result in Chap. 5 and the experimental data (data points) are also shown in Fig. 10.2 for comparison. From Fig. 10.2, we observe that the optimal terminal time is really shorter and the mass of 1,3-PD per unit time at the optimal terminal time is actually higher than previous results.

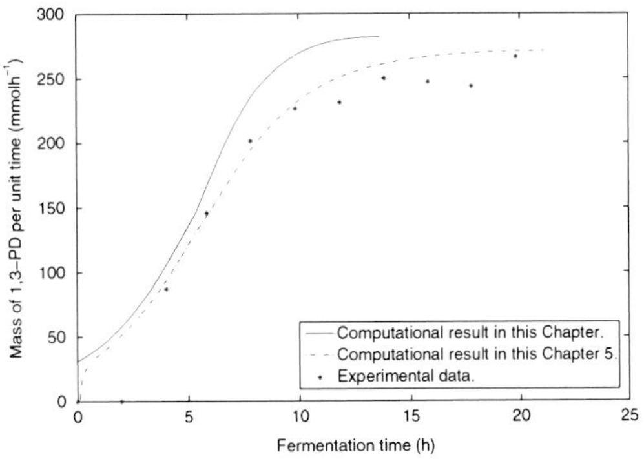


**Fig. 10.2** The mass of 1,3-PD per unit time with respect to fermentation time

## 10.6 Conclusion

In this chapter, we investigated the optimal control of a multistage time-delay system in fed-batch process. We presented the optimal control model and its equivalent form. By the control parameterization technique and the improved DE algorithm, we developed the solution approach to solve the optimal control problem. Numerical results showed the validity of the proposed model and the effectiveness of the developed numerical algorithm.

# Chapter 11
# Optimal Control of Switched Time-Delay Systems

## 11.1 Introduction

In this chapter, we focalize on optimal control of a switched time-delay system arising in constantly fed-batch fermentation process. It is obvious that a model-based efficient approach is necessary to ensure maximum productivity with the lowest possible cost in fed-batch processes, without requiring a human operator. Optimal control of bioprocesses is performed based on precise mathematical models. In view of the existence of time delay and the switching nature in the fed-batch process, a switched time-delay system is proposed to formulate the process. In order to obtain high productivity as well as to reduce the operation costs simultaneously, a free time optimal control model involving the proposed system and subject to continuous state constraints is presented.

Since the presence of free terminal time as well as time delay makes the optimal control problem much more complicated, it should, moreover, be noted that the involving switched time-delay system is highly nonlinear. Therefore it is impossible to obtain analytical solutions of the free time delayed optimal control problem, and one has to resort to numerical solution method. By a time-scaling transformation, we equivalently transcribe the free time delayed optimal control problem into the one with fixed terminal time. Furthermore, using the approach in [115], the switching instants in the resultant delayed optimal control problem are parameterized as a new parameter vector. Subsequently, we develop a numerical solution method for the optimal control problem in two aspects. On the one hand, the smoothing approximation technique is applied to approximate the continuous state constraints by constraints in canonical form. On the other hand, the gradients of the cost functional and constraints with respect to the terminal time and the new parameter vector are derived respectively. On this basis, a gradient-based



C. Liu, Z. Gong, *Optimal Control of Switched Systems Arising in Fermentation Processes*, Springer Optimization and Its Applications 97,
DOI 10.1007/978-3-662-43793-3_11

optimization technique is constructed to seek the optimal control strategy in the free time delayed optimal control problem. Numerical results show effectiveness of the proposed optimization algorithm.

The main reference in this chapter is [147].

## 11.2 Switched Time-Delay Systems

Based on the previous work in Chap. 6 and taking the delay effect on the production of new biomass into account, mass balances of biomass, glycerol, 1,3-PD, and acetate and ethanol in the fed-batch process can be formulated as

$$\begin{cases} \dot{\boldsymbol{x}}(t) = \boldsymbol{f}^i(\boldsymbol{x}(t), \boldsymbol{x}(t-\alpha)), \ t \in (\tau_{i-1}, \tau_i], \ i = 1, 2, \ldots, 2N+1, \\ \boldsymbol{x}(0) = \boldsymbol{x}_0, \\ \boldsymbol{x}(t) = \boldsymbol{\phi}(t), \ t \in [-\bar{\alpha}, 0], \end{cases} \tag{11.1}$$

where $\boldsymbol{x}(t) := (x_1(t), x_2(t), x_3(t), x_4(t), x_5(t), x_6(t))^\top \in \mathbb{R}^6_+$ is the state vector whose components are the concentrations of biomass, substrate, 1,3-PD, and by-products and the volume of culture fluid at $t$ in reactor, respectively. $\alpha$ is a delay argument and is bounded above by a given constant $\bar{\alpha}$ and $\boldsymbol{\phi}(t)$ is a given continuous function on $[-\bar{\alpha}, 0]$. $\tau_i$, $i \in \Lambda := \{0, 1, \ldots, 2N+1\}$, is the switching instant between the batch process and the feed process such that $0 = \tau_0 \leqslant \tau_1 \leqslant \cdots \leqslant \tau_{2N} \leqslant \tau_{2N+1} = T$. Furthermore, for $t \in \left(\tau_{2j}, \tau_{2j+1}\right]$, $j \in \bar{\Lambda}_1 := \{0, 1, \ldots, N\}$,

$$\boldsymbol{f}^{2j+1}(\boldsymbol{x}(t), \boldsymbol{x}(t-\alpha)) = \begin{pmatrix} q_1(\boldsymbol{x}(t))x_1(t-\alpha) \\ -q_2(\boldsymbol{x}(t))x_1(t-\alpha) \\ q_3(\boldsymbol{x}(t))x_1(t-\alpha) \\ q_4(\boldsymbol{x}(t))x_1(t-\alpha) \\ q_5(\boldsymbol{x}(t))x_1(t-\alpha) \\ 0 \end{pmatrix}; \tag{11.2}$$

for $t \in \left(\tau_{2j+1}, \tau_{2j+2}\right]$, $j \in \bar{\Lambda}_2 := \{0, 1, \ldots, N-1\}$,

$$\begin{aligned} &\boldsymbol{f}^{2j+2}(\boldsymbol{x}(t), \boldsymbol{x}(t-\alpha)) \\ &= \begin{pmatrix} q_1(\boldsymbol{x}(t))x_1(t-\alpha) - D(\boldsymbol{x}(t))x_1(t) \\ D(\boldsymbol{x}(t))\left(\dfrac{c_{s0}}{1+r} - x_2(t)\right) - q_2(\boldsymbol{x}(t))x_1(t-\alpha) \\ q_3(\boldsymbol{x}(t))x_1(t-\alpha) - D(\boldsymbol{x}(t))x_3(t) \\ q_4(\boldsymbol{x}(t))x_1(t-\alpha) - D(\boldsymbol{x}(t))x_4(t) \\ q_5(\boldsymbol{x}(t))x_1(t-\alpha) - D(\boldsymbol{x}(t))x_5(t) \\ (1+r)v \end{pmatrix}. \end{aligned} \tag{11.3}$$

In (11.2)–(11.3), $c_{s0} > 0$ denotes the concentration of initial feed of substrate in the medium. $r > 0$ is the velocity ratio of adding alkali to substrate. $v > 0$ is the feeding rate of substrate and is a constant. $D(\boldsymbol{x}(t))$ is the dilution rate defined by

$$D(\boldsymbol{x}(t)) = \frac{(1+r)v}{x_6(t)}. \tag{11.4}$$

The specific growth rate of cells $q_1(\boldsymbol{x}(t))$, the specific consumption rate of substrate $q_2(\boldsymbol{x}(t))$, and the specific formation rates of products $q_\ell(\boldsymbol{x}(t))$, $\ell = 3, 4, 5$, are expressed as

$$q_1(\boldsymbol{x}(t)) = \frac{\Delta_1 x_2(t)}{x_2(t)+k_1} \prod_{\ell=2}^{5} \left(1 - \frac{x_\ell(t)}{x_\ell^*}\right)^{n_\ell}, \tag{11.5}$$

$$q_2(\boldsymbol{x}(t)) = m_2 + q_1(\boldsymbol{x}(t))Y_2 + \frac{\Delta_2 x_2(t)}{x_2(t)+k_2}, \tag{11.6}$$

$$q_\ell(\boldsymbol{x}(t)) = -m_\ell + q_1(\boldsymbol{x}(t))Y_\ell + \frac{\Delta_\ell x_2(t)}{x_2(t)+k_\ell}, \quad \ell = 3, 4, \tag{11.7}$$

$$q_5(\boldsymbol{x}(t)) = q_2(\boldsymbol{x}(t)) \left(\frac{c_1}{c_2 + q_1(\boldsymbol{x}(t))x_2(t)} + \frac{c_3}{c_4 + q_1(\boldsymbol{x}(t))x_2(t)}\right). \tag{11.8}$$

Under anaerobic conditions, the critical concentrations for the cell growth and the values of parameters in (11.5)–(11.8) are as given in Chap. 9.

In the fed-batch process, the switching sequence is preassigned and the switching instants $\tau_i$, $i = 1, 2, \ldots, 2N$, together with the terminal time $T$ are decision variables. Nevertheless, since biological considerations limit the rate of switching, there are maximal and minimal time durations that are spent on each of the batch and feed processes. On this basis, define the set of admissible switching instants and terminal time as

$$\Gamma := \{(\tau_1, \tau_2, \ldots, \tau_{2N+1})^\top \in \mathbb{R}^{2N+1} \mid \rho_i \leqslant \tau_i - \tau_{i-1} \leqslant \varrho_i,$$
$$i = 1, 2, \ldots, 2N+1\}, \tag{11.9}$$

where $\rho_j$ and $\varrho_j$ are the minimal and the maximal time durations, respectively. Accordingly, any $\boldsymbol{\tau} \in \Gamma$ is regarded as an admissible vector of switching instants and terminal time.

There exist critical concentrations of biomass, substrate, and products and the volume of culture fluid, outside which cells cease to grow. Hence, it is biologically meaningful to restrict the concentrations of biomass, substrate, and products and the volume of culture fluid within a set $\tilde{W}$ defined as

$$\boldsymbol{x}(t) \in \tilde{W} := \prod_{\ell=1}^{6} [x_{*\ell}, x_\ell^*], \ \forall t \in [0, T], \tag{11.10}$$

where $x_{*\ell}, x_\ell^*$, $\ell = 1, 2, \ldots, 5$, are as given in Table 5.1, $x_{*6} = 4$ and $x_6^* = 7$.

For the system (11.1), some important properties are given in the following theorems.

**Theorem 11.1.** *The functions* $\boldsymbol{f}^i(\cdot,\cdot)$, $i = 1, 2, \ldots, 2N+1$, *defined in* (11.2) *and* (11.3) *satisfy the following conditions:*

(a) $\boldsymbol{f}^i(\cdot,\cdot) : \mathbb{R}^6_+ \times \mathbb{R}^6_+ \to \mathbb{R}^6$, *together with its partial derivatives with respect to* $\boldsymbol{x}$ *and* $\boldsymbol{y}$, *are continuous on* $\mathbb{R}^6_+ \times \mathbb{R}^6_+$.
(b) *There exists a constant* $K > 0$ *such that*

$$\|\boldsymbol{f}^i(\boldsymbol{x},\boldsymbol{y})\| \leqslant K(1 + \|\boldsymbol{x}\| + \|\boldsymbol{y}\|), \forall (\boldsymbol{x},\boldsymbol{y}) \in \mathbb{R}^6_+ \times \mathbb{R}^6_+, \tag{11.11}$$

*where* $\|\cdot\|$ *denotes the Euclidean norm.*

*Proof.* (a) This conclusion can be obtained by the expression of $\boldsymbol{f}^i$ in (11.2) and (11.3).
(b) The proof of this theorem is similar to the proof that is given for Theorem 6.1 in Chap. 6. □

**Theorem 11.2.** *For each* $\boldsymbol{\tau} \in \Gamma$, *the system* (11.1) *has a unique continuous solution on* $[-\bar{\alpha}, T]$, *denoted by* $\boldsymbol{x}(\cdot|\boldsymbol{\tau})$. *Furthermore,* $\boldsymbol{x}(\cdot|\boldsymbol{\tau})$ *satisfies that*

$$\boldsymbol{x}(t|\boldsymbol{\tau}) = \boldsymbol{x}(\tau_{i-1}|\boldsymbol{\tau}) + \int_{\tau_{i-1}}^{t} \boldsymbol{f}^i(\boldsymbol{x}(s|\boldsymbol{\tau}), \boldsymbol{x}(s-\alpha|\boldsymbol{\tau}))\mathrm{d}s,$$
$$\forall t \in (\tau_{i-1}, \tau_i],\ i = 1, 2, \ldots, 2N+1, \tag{11.12}$$

*and* $\boldsymbol{x}(t|\boldsymbol{\tau}) = \boldsymbol{\phi}(t)$, $\forall t \in [-\bar{\alpha}, 0]$.

*Proof.* The proof can be obtained by Theorem 11.1 and the theory of delay-differential equations [95]. □

**Theorem 11.3.** *Given the continuous function* $\boldsymbol{\phi}(t)$ *and the initial state* $\boldsymbol{x}_0$, *the unique solution* $\boldsymbol{x}(\cdot|\boldsymbol{\tau})$ *of the system* (11.1) *is uniformly bounded.*

*Proof.* Since $\boldsymbol{\phi}(t)$ is continuous on $[-\bar{\alpha}, 0]$, there exists a real number $0 \leqslant M < +\infty$ such that

$$\sup\{\|\boldsymbol{\phi}(t)\| \mid t \in [-\bar{\alpha}, 0]\} \leqslant M.$$

Thus,

$$\|\boldsymbol{x}(t|\boldsymbol{\tau})\| \leqslant M,\ \forall t \in [-\bar{\alpha}, 0].$$

In view of Theorems 11.1 and 11.2, we obtain that

$$\|\boldsymbol{x}(t|\boldsymbol{\tau})\| \leqslant \|\boldsymbol{x}_0\| + \sum_{k=1}^{i-1} \int_{\tau_{k-1}}^{\tau_k} \|\boldsymbol{f}^k(\boldsymbol{x}(s|\boldsymbol{\tau}), \boldsymbol{x}(s-\alpha|\boldsymbol{\tau}))\|\mathrm{d}s$$
$$+ \int_{\tau_{i-1}}^{t} \|\boldsymbol{f}^i(\boldsymbol{x}(s|\boldsymbol{\tau}), \boldsymbol{x}(s-\alpha|\boldsymbol{\tau}))\|\mathrm{d}s,$$

$$\leqslant \|x_0\| + \int_0^t K(1 + \|x(s|\tau)\| + \|x(s-\alpha|\tau)\|)\mathrm{d}s,$$

$$\leqslant M + K\bar{\alpha}M + K\int_0^t (1 + 2\|x(s|\tau)\|)\mathrm{d}s, \quad \forall t \in (0, T].$$

By Lemma 4.1, it follows that

$$\|x(t|\tau)\| \leqslant (M + K\bar{\alpha}M + KT_{\max})\exp(2KT_{\max}), \quad \forall t \in (0, T],$$

in which $T_{\max} := \sum_{i=1}^{2N+1} \varrho_i$. Therefore,

$$\|x(t|\tau)\| \leqslant M', \quad \forall t \in [-\bar{\alpha}, T],$$

where $M' := \max\{M, (M + K\bar{\alpha}M + KT_{\max})\exp(2KT_{\max})\}$. □

## 11.3 Optimal Control Problems

In fermentation process, it is desired to maximize the target product at the end of the process as well as to minimize the operation costs simultaneously by operating some control variables. In this section, the optimal control problem and its equivalent form in constantly fed-batch process are discussed.

### *11.3.1 Free Time Delayed Optimal Control Problem*

For mathematical convenience, define the set of the solutions to the system (11.1) as

$$\mathcal{S}_0 = \{x(\cdot|\tau)|\ x(\cdot|\tau) \text{ is the solution of the system (11.1)}\}. \tag{11.13}$$

Due to the constraint (11.10) on the state, we can define the feasible set of the solutions

$$\mathcal{S} = \left\{x(\cdot|\tau)|\ x(\cdot|\tau) \in \mathcal{S}_0 \text{ and } x(t|\tau) \in \tilde{W}, \forall t \in [0, T]\right\}. \tag{11.14}$$

Consequently, the feasible set of the switching instants and the terminal time is

$$F = \{\tau \in \Gamma|\ x(\cdot|\tau) \in \mathcal{S}\}. \tag{11.15}$$

Since the mass of the 1,3-PD and the duration of the process are two key elements to affect the profit in fermentation process, the mass of 1,3-PD per unit time at the terminal time is taken as the cost functional which is different from the one in Chap. 6. As a result, the free time delayed optimal control problem (FDOC) in fed-batch process can be formulated as

$$\text{(FDOC)} \qquad \min J(\boldsymbol{\tau}) = -\frac{x_3(T|\boldsymbol{\tau})x_6(T|\boldsymbol{\tau})}{T}$$
$$\text{s.t. } \boldsymbol{\tau} \in F,$$

where $x_3(T|\boldsymbol{\tau})$ and $x_6(T|\boldsymbol{\tau})$ are, respectively, the third and the sixth components of the solution to the system (11.1) at the terminal time $T$.

*Remark 11.1.* Note that the (FDOC) is of nonstandard feature because the terminal time as well as the switching instants is the variable to be determined. Thus, the (FDOC) is actually a constrained free time delayed optimal control problem.

### *11.3.2 The Equivalent Optimal Control Problem*

It is difficult to solve the (FDOC) using the existing numerical techniques. The main difficulty is the implicit dependence of the system state on the terminal time. To surmount this difficulty, we transcribe the (FDOC) into an equivalent optimal control problem with fixed terminal time.

First of all, the time-scaling transformation from $t \in [0, T]$ to $s \in [0, 1]$ can be established as follows:

$$t = Ts. \tag{11.16}$$

Moreover, let $\tilde{\boldsymbol{x}}(s) := \boldsymbol{x}(t(s))$, $\tilde{\alpha} := \frac{\alpha}{T}$, $\bar{\tilde{\alpha}} := \frac{\bar{\alpha}}{T}$, $\boldsymbol{g}^i(\tilde{\boldsymbol{x}}(s), \tilde{\boldsymbol{x}}(s-\tilde{\alpha}), T) := T\boldsymbol{f}^i(\tilde{\boldsymbol{x}}(s), \tilde{\boldsymbol{x}}(s-\tilde{\alpha}))$, $\tilde{\boldsymbol{\phi}}(s) := \boldsymbol{\phi}(t(s))$ and $s_i := \frac{\tau_i}{T}$, $i = 1, 2, \ldots, 2N+1$. Then, the original switched time-delay system (11.1) takes the form

$$\begin{cases} \dot{\tilde{\boldsymbol{x}}}(s) = \boldsymbol{g}^i(\tilde{\boldsymbol{x}}(s), \tilde{\boldsymbol{x}}(s-\tilde{\alpha}), T),\ s \in (s_{i-1}, s_i],\ i = 1, 2, \ldots, 2N+1, \\ \tilde{\boldsymbol{x}}(0) = \boldsymbol{x}_0, \\ \tilde{\boldsymbol{x}}(s) = \tilde{\boldsymbol{\phi}}(s), s \in [-\bar{\tilde{\alpha}}, 0]. \end{cases} \tag{11.17}$$

Furthermore, let

$$\xi_i := s_i - s_{i-1},\ i \in \Lambda, \tag{11.18}$$

be the duration between $s_{i-1}$ and $s_i$. Clearly,

$$s_i = \sum_{k=1}^{i} \xi_k,\ i = 1, 2, \ldots, 2N + 1. \tag{11.19}$$

Let $\boldsymbol{\xi} := (\xi_1, \xi_2, \ldots, \xi_{2N+1}) \in \mathbb{R}^{2N+1}$ be the duration vector. It is obvious that

$$\xi_i \geqslant 0,\ i = 1, 2, \ldots, 2N + 1, \tag{11.20}$$

and

$$\sum_{i=1}^{2N+1} \xi_i = 1. \tag{11.21}$$

With this notation, we note that the determination of the switching instants is equivalent to the determination of the duration vector. As a result, $\tilde{\boldsymbol{x}}$ can be viewed as a implicit function of the terminal time and the duration vector, i.e.,

$$\tilde{\boldsymbol{x}}(s) = \tilde{\boldsymbol{x}}(s|T, \xi_{i-1}, \xi_{i-2}, \ldots, \xi_1), \tag{11.22}$$

for $s \in (s_{i-1}, s_i]$, $i = 1, 2, \ldots, 2N + 1$. Then, the system (11.17) can be written as

$$\begin{aligned} &\frac{\partial \tilde{\boldsymbol{x}}(s|T, \xi_{i-1}, \xi_{i-2}, \ldots, \xi_1)}{\partial s} \\ &\quad = \boldsymbol{g}^i(\tilde{\boldsymbol{x}}(s|T, \xi_{i-1}, \xi_{i-2}, \ldots, \xi_1), \tilde{\boldsymbol{x}}(s - \tilde{\alpha}|T, \xi_{i-1}, \xi_{i-2}, \ldots, \xi_1), T), \\ &\qquad\qquad s \in (s_{i-1}, s_i], \end{aligned} \tag{11.23}$$

with intermediate condition

$$\begin{aligned} \tilde{\boldsymbol{x}}(s|T, \xi_{i-1}, \xi_{i-2}, \ldots, \xi_1)|_{s=s_{i-1}} &= \tilde{\boldsymbol{x}}(s|T, \xi_{i-2}, \xi_{i-3}, \ldots, \xi_1)|_{s=s_{i-1}}, \\ &\qquad i = 1, 2, \ldots, 2N + 1, \end{aligned} \tag{11.24}$$

and the initial conditions

$$\tilde{\boldsymbol{x}}(s)|_{s=0} = \boldsymbol{x}_0, \tag{11.25}$$

$$\tilde{\boldsymbol{x}}(s) = \tilde{\boldsymbol{\phi}}(s),\ s \in [-\bar{\tilde{\alpha}}, 0]. \tag{11.26}$$

Consequently, the constraints (11.9) and (11.10) become, respectively,

$$\begin{aligned} \Xi := \{(T, \xi_1, \xi_2, \ldots, \xi_{2N+1}) \in \mathbb{R}^{2N+2} |\ &\tilde{\rho}_i \leqslant \xi_i \leqslant \tilde{\varrho}_i, i = 1, 2, \ldots, 2N + 1, \\ &\text{and} \sum_{i=1}^{2N+1} \tilde{\rho}_i \leqslant T \leqslant \sum_{i=1}^{2N+1} \tilde{\varrho}_i\}, \end{aligned} \tag{11.27}$$

and

$$\tilde{\boldsymbol{x}}(s|T,\xi) \in \tilde{W}, \ s \in [0,1], \tag{11.28}$$

where $\tilde{\rho}_i = \dfrac{\rho_i}{T}$ and $\tilde{\varrho}_i = \dfrac{\varrho_i}{T}$. Furthermore, the feasible set of the terminal time and the duration vector can be rewritten as

$$\tilde{F} = \left\{(T,\boldsymbol{\xi}) \in \Xi |\ \tilde{\boldsymbol{x}}(s|T,\boldsymbol{\xi}) \in \tilde{W}, \forall s \in [0,1]\right\}. \tag{11.29}$$

Now, we can establish the equivalently optimal control problem (EOC) of (FDOC) as follows:

$$\text{(EOC)} \qquad \min \tilde{J}(T,\boldsymbol{\xi}) = -\frac{\tilde{x}_3(1|T,\boldsymbol{\xi})\tilde{x}_6(1|T,\boldsymbol{\xi})}{T}$$

$$\text{s.t.}\ \ (T,\boldsymbol{\xi}) \in \tilde{F}.$$

## 11.4 Numerical Solution Methods

In this section, we shall develop a numerical solution method to solve the (FDOC).

### *11.4.1 Approximation Problem*

The (EOC) is essentially an optimization problem with continuous state inequality constraint (11.28). This type of constraints often arises in the actual process. We use the method in Chap. 6 to handle this type of constraints. To begin with, let

$$h_\ell(\tilde{\boldsymbol{x}}(s|T,\boldsymbol{\xi})) = x_\ell^* - \tilde{x}_\ell(s|T,\boldsymbol{\xi}),$$
$$h_{6+\ell}(\tilde{\boldsymbol{x}}(s|T,\boldsymbol{\xi})) = \tilde{x}_\ell(s|T,\boldsymbol{\xi}) - x_{*\ell},\ \ell = 1,2,\ldots,6.$$

Then, the constraint (11.28) can be equivalently transcribed into

$$G(T,\boldsymbol{\xi}) = 0, \tag{11.30}$$

where $G(T,\boldsymbol{\xi}) := \sum_{l=1}^{12}\int_0^1 \min\{0, h_l(\tilde{\boldsymbol{x}}(s|T,\boldsymbol{\xi}))\}\mathrm{d}s$. However, since $G(\cdot,\cdot)$ is non-differentiable at the point $h_l = 0$, standard optimization routines would have difficulties in dealing with this type of equality constraints. Thus, we replace (11.30) with

$$\tilde{G}_{\varepsilon,\gamma}(T,\boldsymbol{\xi}) := \gamma + \sum_{l=1}^{12}\int_0^1 \varphi_\varepsilon(h_l(\tilde{\boldsymbol{x}}(s|T,\boldsymbol{\xi})))\mathrm{d}s \geqslant 0, \tag{11.31}$$

where $\varepsilon > 0$, $\gamma > 0$ and

$$\varphi_\varepsilon(\eta) = \begin{cases} \eta, & \text{if } \eta < -\varepsilon, \\ -\dfrac{(\eta - \varepsilon)^2}{4\varepsilon}, & \text{if } -\varepsilon \leqslant \eta \leqslant \varepsilon, \\ 0, & \text{if } \eta > \varepsilon. \end{cases} \tag{11.32}$$

Note that $\tilde{G}_{\varepsilon,\gamma}(T, \boldsymbol{\xi})$ is smooth in $T$ and $\boldsymbol{\xi}$. In addition, we can define the corresponding approximately feasible set of the terminal time and the duration vector as

$$\tilde{F}_{\varepsilon,\gamma} = \left\{(T, \boldsymbol{\xi}) \in \Xi \mid \tilde{G}_{\varepsilon,\gamma}(T, \boldsymbol{\xi}) \geqslant 0\right\}. \tag{11.33}$$

Then, the (EOC) can be approximated by the approximately optimal control problem as follows:

$$\left(\mathrm{EOC}_{\varepsilon,\gamma}\right) \qquad \min \tilde{J}_{\varepsilon,\gamma}(T, \boldsymbol{\xi}) = -\frac{\tilde{x}_3(1|T, \boldsymbol{\xi})\tilde{x}_6(1|T, \boldsymbol{\xi})}{T} \tag{11.34}$$

$$\text{s.t. } (T, \boldsymbol{\xi}) \in \tilde{F}_{\varepsilon,\gamma}.$$

## 11.4.2 A Computational Procedure

To solve the (FDOC), we need to solve a sequence of problems $\left\{\left(\mathrm{EOC}_{\varepsilon,\gamma}\right)\right\}$. Each $\left(\mathrm{EOC}_{\varepsilon,\gamma}\right)$ can be solved as a mathematical programming problem using any efficient optimization technique, such as the sequential quadratic programming (SQP) [189]. For this, we need the gradients of the cost functional and the constraints with respect to the terminal time and the duration vector.

Define

$$\bar{\boldsymbol{\chi}}(s) := \begin{cases} \dot{\bar{\boldsymbol{\phi}}}(s), & \text{if } s \in \left[-\bar{\tilde{\alpha}}, 0\right], \\ \boldsymbol{g}^i\left(\tilde{\boldsymbol{x}}(s), \tilde{\boldsymbol{x}}(s - \tilde{\alpha}), T\right), & \text{if } s \in (s_{i-1}, s_i] \text{ for some } i \in \Lambda. \end{cases}$$

These gradients are given in the following theorems.

**Theorem 11.4.** *For each $\varepsilon > 0$ and $\gamma > 0$, the gradient of the cost functional* (11.34) *with respect to the terminal time $T$ is given by*

$$\frac{\partial \tilde{J}_{\varepsilon,\gamma}(T, \boldsymbol{\xi})}{\partial T} = -\frac{\psi_3(1|T, \boldsymbol{\xi})\tilde{x}_6(1|T, \boldsymbol{\xi})T + \tilde{x}_3(1|T, \boldsymbol{\xi})\psi_6(1|T, \boldsymbol{\xi})T - \tilde{x}_3(1|T, \boldsymbol{\xi})\tilde{x}_6(1|T, \boldsymbol{\xi})}{T^2}, \tag{11.35}$$

*where $\boldsymbol{\psi}(s)$ is the solution of the following auxiliary time-delay system:*

$$\begin{aligned}\dot{\boldsymbol{\psi}}(s) = & \frac{\partial \boldsymbol{g}^i(\tilde{\boldsymbol{x}}(s|T,\xi_{i-1},\ldots,\xi_1),\tilde{\boldsymbol{x}}(s-\tilde{\alpha}|T,\xi_{i-1},\ldots,\xi_1),T)}{\partial \tilde{\boldsymbol{x}}(s)}\boldsymbol{\psi}(s) \\ & + \frac{\partial \boldsymbol{g}^i(\tilde{\boldsymbol{x}}(s|T,\xi_{i-1},\ldots,\xi_1),\tilde{\boldsymbol{x}}(s-\tilde{\alpha}|T,\xi_{i-1},\ldots,\xi_1),T)}{\partial \tilde{\boldsymbol{x}}(s-\tilde{\alpha})}\boldsymbol{\psi}(s-\tilde{\alpha}) \\ & + \frac{\tilde{\alpha}}{T}\frac{\partial \boldsymbol{g}^i(\tilde{\boldsymbol{x}}(s|T,\xi_{i-1},\ldots,\xi_1),\tilde{\boldsymbol{x}}(s-\tilde{\alpha}|T,\xi_{i-1},\ldots,\xi_1),T)}{\partial \tilde{\boldsymbol{x}}(s-\tilde{\alpha})}\bar{\boldsymbol{\chi}}(s-\tilde{\alpha}) \\ & + \boldsymbol{f}^i(\tilde{\boldsymbol{x}}(s|T,\xi_{i-1},\ldots,\xi_1),\tilde{\boldsymbol{x}}(s-\tilde{\alpha}|T,\xi_{i-1},\ldots,\xi_1)), \\ & \qquad \forall s \in (s_{i-1},s_i],\ i = 1,2,\ldots,2N+1, \end{aligned} \tag{11.36}$$

*with*

$$\boldsymbol{\psi}(0) = \boldsymbol{0}, \tag{11.37}$$

$$\boldsymbol{\psi}(s) = \frac{\partial \tilde{\boldsymbol{\phi}}(s)}{\partial T},\ \forall s \in \left[-\bar{\tilde{\alpha}},0\right]. \tag{11.38}$$

*Proof.* For calculating the gradient of the cost functional $\tilde{J}_{\varepsilon,\gamma}(T,\boldsymbol{\xi})$ with respect to $T$, we need to calculate $\dfrac{\partial \tilde{\boldsymbol{x}}(1|T,\boldsymbol{\xi})}{\partial T}$. Note that $\boldsymbol{g}^i(\tilde{\boldsymbol{x}}(s),\tilde{\boldsymbol{x}}(s-\tilde{\alpha}),T)$, $i = 1,2,\ldots,2N+1$, are continuous differentiable with respect to their arguments. Therefore, by taking the partial differentiation of both sides (11.23) with respect to $T$, we obtain

$$\begin{aligned} & \frac{\partial^2 \tilde{\boldsymbol{x}}(s|T,\xi_{i-1},\ldots,\xi_1)}{\partial T\,\partial s} \\ & = \frac{\partial \boldsymbol{g}^i(\tilde{\boldsymbol{x}}(s|T,\xi_{i-1},\ldots,\xi_1),\tilde{\boldsymbol{x}}(s-\tilde{\alpha}|T,\xi_{i-1},\ldots,\xi_1),T)}{\partial \tilde{\boldsymbol{x}}(s)}\frac{\partial \tilde{\boldsymbol{x}}(s|T,\xi_{i-1},\ldots,\xi_1)}{\partial T} \\ & \quad + \frac{\partial \boldsymbol{g}^i(\tilde{\boldsymbol{x}}(s|T,\xi_{i-1},\ldots,\xi_1),\tilde{\boldsymbol{x}}(s-\tilde{\alpha}|T,\xi_{i-1},\ldots,\xi_1),T)}{\partial \tilde{\boldsymbol{x}}(s-\tilde{\alpha})}\frac{\partial \tilde{\boldsymbol{x}}(s-\tilde{\alpha}|T,\xi_{i-1},\ldots,\xi_1)}{\partial T} \\ & \quad + \frac{\tilde{\alpha}}{T}\frac{\partial \boldsymbol{g}^i(\tilde{\boldsymbol{x}}(s|T,\xi_{i-1},\ldots,\xi_1),\tilde{\boldsymbol{x}}(s-\tilde{\alpha}|T,\xi_{i-1},\ldots,\xi_1),T)}{\partial \tilde{\boldsymbol{x}}(s-\tilde{\alpha})}\bar{\boldsymbol{\chi}}(s-\tilde{\alpha}) \\ & \quad + \boldsymbol{f}^i(\tilde{\boldsymbol{x}}(s|T,\xi_{i-1},\ldots,\xi_1),\tilde{\boldsymbol{x}}(s-\tilde{\alpha}|T,\xi_{i-1},\ldots,\xi_1)), \\ & \qquad \forall s \in (s_{i-1},s_i],\ i = 1,2,\ldots,2N+1. \end{aligned}$$

Moreover,

$$\frac{\partial \tilde{\boldsymbol{x}}(0)}{\partial T} = \frac{\partial}{\partial T}\left\{\boldsymbol{x}_0\right\} = \boldsymbol{0},$$

$$\frac{\partial \tilde{\boldsymbol{x}}(s)}{\partial T} = \frac{\partial \tilde{\boldsymbol{\phi}}(s)}{\partial T}, \ \forall s \in \left[-\bar{\bar{\alpha}}, 0\right].$$

Using the fact that

$$\frac{\partial^2 \tilde{\boldsymbol{x}}(s|T, \xi_{i-1}, \ldots, \xi_1)}{\partial T \partial s} = \frac{\partial}{\partial s}\left(\frac{\partial \tilde{\boldsymbol{x}}(s|T, \xi_{i-1}, \ldots, \xi_1)}{\partial T}\right)$$

and defining

$$\boldsymbol{\psi}(s|T, \boldsymbol{\xi}) := \frac{\partial \tilde{\boldsymbol{x}}(s|T, \xi_{i-1}, \ldots, \xi_1)}{\partial T},$$

we obtain the conclusion (11.35). The proof is complete. □

**Theorem 11.5.** *For each $\varepsilon > 0$ and $\gamma > 0$, the gradient of the cost functional* (11.34) *with respect to the duration vector $\boldsymbol{\xi}$ is given by*

$$\frac{\partial \tilde{J}_{\varepsilon,\gamma}(T, \boldsymbol{\xi})}{\partial \xi_i} = -\frac{\varphi_3^i(1|T, \boldsymbol{\xi})\tilde{x}_6(1|T, \boldsymbol{\xi}) + \tilde{x}_3(1|T, \boldsymbol{\xi})\varphi_6^i(1|T, \boldsymbol{\xi})}{T},$$
$$i = 1, 2, \ldots, 2N, \tag{11.39}$$

*where $\boldsymbol{\varphi}^i(s)$ are the solution of the following auxiliary time-delay systems:*

$$\dot{\boldsymbol{\varphi}}^i(s) = \frac{\partial \boldsymbol{g}^{i+1}(\tilde{\boldsymbol{x}}(s|T, \xi_i, \ldots, \xi_1), \tilde{\boldsymbol{x}}(s-\tilde{\alpha}|T, \xi_i, \ldots, \xi_1), T)}{\partial \tilde{\boldsymbol{x}}(s)}\boldsymbol{\varphi}^i(s)$$
$$+ \frac{\partial \boldsymbol{g}^{i+1}(\tilde{\boldsymbol{x}}(s|T, \xi_i, \ldots, \xi_1), \tilde{\boldsymbol{x}}(s-\tilde{\alpha}|T, \xi_i, \ldots, \xi_1), T)}{\partial \tilde{\boldsymbol{x}}(s-\tilde{\alpha})}\boldsymbol{\varphi}^i(s-\tilde{\alpha}),$$
$$s \in (s_i, s_{i+1}],$$

$$\ldots$$

$$\dot{\boldsymbol{\varphi}}^i(s) = \frac{\partial \boldsymbol{g}^{2N+1}(\tilde{\boldsymbol{x}}(s|T, \xi_{2N}, \ldots, \xi_1), \tilde{\boldsymbol{x}}(s-\tilde{\alpha}|T, \xi_{2N}, \ldots, \xi_1), T)}{\partial \tilde{\boldsymbol{x}}(s)}\boldsymbol{\varphi}^i(s)$$
$$+ \frac{\partial \boldsymbol{g}^{2N+1}(\tilde{\boldsymbol{x}}(s|T, \xi_{2N}, \ldots, \xi_1), \tilde{\boldsymbol{x}}(s-\tilde{\alpha}|T, \xi_{2N}, \ldots, \xi_1), T)}{\partial \tilde{\boldsymbol{x}}(s-\tilde{\alpha})}\boldsymbol{\varphi}^i(s-\tilde{\alpha}),$$
$$s \in (s_{2N}, 1], \tag{11.40}$$

*with*

$$\boldsymbol{\varphi}^i(s_i) = \boldsymbol{g}^i(\tilde{\boldsymbol{x}}(s_i|T, \xi_{i-1}, \ldots, \xi_1), \tilde{\boldsymbol{x}}(s_i-\tilde{\alpha}|T, \xi_{i-1}, \ldots, \xi_1), T), \tag{11.41}$$

$$\boldsymbol{\varphi}^i(s) = \boldsymbol{0}, \forall s \in \left[-\bar{\bar{\alpha}}, s_i\right]. \tag{11.42}$$

*Furthermore,*

$$\frac{\partial \tilde{J}_{\varepsilon,\gamma}(T,\boldsymbol{\xi})}{\partial \xi_{2N+1}} = -\frac{\varphi_3^{2N+1}(1|T,\boldsymbol{\xi})\tilde{x}_6(1|T,\boldsymbol{\xi}) + \tilde{x}_3(1|T,\boldsymbol{\xi})\varphi_6^{2N+1}(1|T,\boldsymbol{\xi})}{T}, \tag{11.43}$$

*where*

$$\boldsymbol{\varphi}^{2N+1}(s) = \boldsymbol{g}^{2N+1}(\tilde{\boldsymbol{x}}(s|T,\xi_{2N},\ldots,\xi_1), \tilde{\boldsymbol{x}}(s-\tilde{\alpha}|T,\xi_{2N},\ldots,\xi_1), T),$$
$$s \in (s_{2N}, 1], \tag{11.44}$$

*with*

$$\boldsymbol{\varphi}^{2N+1}(s) = \mathbf{0},\ s \in \left[-\bar{\tilde{\alpha}}, s_{2N}\right]. \tag{11.45}$$

*Proof.* For $i = 1,2,\ldots,2N$, by taking the partial differentiation of both sides (11.23) with respect to $\xi_i$, we obtain

$$\begin{aligned}
&\frac{\partial^2 \tilde{\boldsymbol{x}}(s|T,\xi_{i-1},\ldots,\xi_1)}{\partial \xi_i\, \partial s} \\
&= \frac{\partial \boldsymbol{g}^i(\tilde{\boldsymbol{x}}(s|T,\xi_{i-1},\ldots,\xi_1), \tilde{\boldsymbol{x}}(s-\tilde{\alpha}|T,\xi_{i-1},\ldots,\xi_1), T)}{\partial \tilde{\boldsymbol{x}}(s)} \frac{\partial \tilde{\boldsymbol{x}}(s|T,\xi_{i-1},\ldots,\xi_1)}{\partial \xi_i} \\
&\quad + \frac{\partial \boldsymbol{g}^i(\tilde{\boldsymbol{x}}(s|T,\xi_{i-1},\ldots,\xi_1), \tilde{\boldsymbol{x}}(s-\tilde{\alpha}|T,\xi_{i-1},\ldots,\xi_1), T)}{\partial \tilde{\boldsymbol{x}}(s-\tilde{\alpha})} \frac{\partial \tilde{\boldsymbol{x}}(s-\tilde{\alpha}|T,\xi_{i-1},\ldots,\xi_1)}{\partial \xi_i}.
\end{aligned}$$

Since $\tilde{\boldsymbol{x}}(s)$ is only dependent on those $\xi_j$ such that $\sum_{j=1}^{i} \xi_j \leqslant s$, it follows that

$$\frac{\partial \boldsymbol{x}(s|T,\xi_j,\ldots,\xi_1)}{\partial \xi_j} = \mathbf{0},\ s \leqslant \sum_{k=1}^{j} \xi_k, i > k.$$

Let

$$\boldsymbol{\varphi}^i(s|T,\xi_k,\ldots,\xi_1) := \frac{\partial \tilde{\boldsymbol{x}}(s|\xi_k,\ldots,\xi_1)}{\partial \xi_i},\ k = 1,2,\ldots,2N,$$

and we obtain the conclusion (11.39). The gradient (11.43) can be derived similarly. The proof is complete. □

**Theorem 11.6.** *For each $\varepsilon > 0$ and $\gamma > 0$, the constraint* (11.31) *with respect to the terminal time $T$ is given by*

$$\frac{\partial \tilde{G}_{\varepsilon,\gamma}(T,\boldsymbol{\xi})}{\partial T} = \sum_{l=1}^{12} \int_0^1 \frac{\partial \varphi_\varepsilon(h_l(\tilde{\boldsymbol{x}}(s|T,\boldsymbol{\xi})))}{\partial h_l} \frac{\partial h_l(\tilde{\boldsymbol{x}}(s|T,\boldsymbol{\xi}))}{\partial \tilde{\boldsymbol{x}}} \boldsymbol{\psi}(s)\mathrm{d}s, \quad (11.46)$$

*where* $\boldsymbol{\psi}(s)$ *is the solution of the following auxiliary time-delay system:*

$$\begin{aligned}
\dot{\boldsymbol{\psi}}(s) = {} & \frac{\partial \boldsymbol{g}^i(\tilde{\boldsymbol{x}}(s|T,\xi_{i-1},\ldots,\xi_1),\tilde{\boldsymbol{x}}(s-\tilde{\alpha}|T,\xi_{i-1},\ldots,\xi_1),T)}{\partial \tilde{\boldsymbol{x}}(s)} \boldsymbol{\psi}(s) \\
& + \frac{\partial \boldsymbol{g}^i(\tilde{\boldsymbol{x}}(s|T,\xi_{i-1},\ldots,\xi_1),\tilde{\boldsymbol{x}}(s-\tilde{\alpha}|T,\xi_{i-1},\ldots,\xi_1),T)}{\partial \tilde{\boldsymbol{x}}(s-\tilde{\alpha})} \boldsymbol{\psi}(s-\tilde{\alpha}) \\
& + \frac{\tilde{\alpha}}{T} \frac{\partial \boldsymbol{g}^i(\tilde{\boldsymbol{x}}(s|T,\xi_{i-1},\ldots,\xi_1),\tilde{\boldsymbol{x}}(s-\tilde{\alpha}|T,\xi_{i-1},\ldots,\xi_1),T)}{\partial \tilde{\boldsymbol{x}}(s-\tilde{\alpha})} \bar{\boldsymbol{\chi}}(s-\tilde{\alpha}) \\
& + \boldsymbol{f}^i(\tilde{\boldsymbol{x}}(s|T,\xi_{i-1},\ldots,\xi_1),\tilde{\boldsymbol{x}}(s-\tilde{\alpha}|T,\xi_{i-1},\ldots,\xi_1)), \\
& \forall s \in (s_{i-1},s_i],\ i = 1,2,\ldots,2N+1, \quad (11.47)
\end{aligned}$$

*with*

$$\boldsymbol{\psi}(0) = \boldsymbol{0}, \quad (11.48)$$

$$\boldsymbol{\psi}(s) = \frac{\partial \tilde{\boldsymbol{\phi}}(s)}{\partial T}, \quad \forall s \in \left[-\bar{\tilde{\alpha}}, 0\right]. \quad (11.49)$$

*Proof.* We can complete the proof using a method similar to the proof of Theorem 11.4. □

**Theorem 11.7.** *For each* $\varepsilon > 0$ *and* $\gamma > 0$, *the gradient of the constraint* (11.31) *with respect to the duration vector* $\boldsymbol{\xi}$ *is given by*

$$\frac{\partial \tilde{G}_{\varepsilon,\gamma}(T,\boldsymbol{\xi})}{\partial \xi_i} = \sum_{l=1}^{12} \int_0^1 \frac{\partial \varphi_\varepsilon(h_l(\tilde{\boldsymbol{x}}(s|T,\boldsymbol{\xi})))}{\partial h_l} \frac{\partial h_l(\tilde{\boldsymbol{x}}(s|T,\boldsymbol{\xi}))}{\partial \tilde{\boldsymbol{x}}} \boldsymbol{\varphi}^i(s)\mathrm{d}s,$$
$$i = 1,2,\ldots,2N, \quad (11.50)$$

*where* $\boldsymbol{\varphi}^i(s)$ *are the solution of the following auxiliary time-delay systems:*

$$\begin{aligned}
\dot{\boldsymbol{\varphi}}^i(s) = {} & \frac{\partial \boldsymbol{g}^{i+1}(\tilde{\boldsymbol{x}}(s|T,\xi_i,\ldots,\xi_1),\tilde{\boldsymbol{x}}(s-\tilde{\alpha}|T,\xi_i,\ldots,\xi_1),T)}{\partial \tilde{\boldsymbol{x}}(s)} \boldsymbol{\varphi}^i(s) \\
& + \frac{\partial \boldsymbol{g}^{i+1}(\tilde{\boldsymbol{x}}(s|T,\xi_i,\ldots,\xi_1),\tilde{\boldsymbol{x}}(s-\tilde{\alpha}|T,\xi_i,\ldots,\xi_1),T)}{\partial \tilde{\boldsymbol{x}}(s-\tilde{\alpha})} \boldsymbol{\varphi}^i(s-\tilde{\alpha}), \\
& s \in (s_i, s_{i+1}], \\
& \ldots
\end{aligned}$$

$$\dot{\boldsymbol{\varphi}}^i(s) = \frac{\partial \boldsymbol{g}^{2N+1}(\tilde{\boldsymbol{x}}(s|T,\xi_{2N},\ldots,\xi_1),\tilde{\boldsymbol{x}}(s-\tilde{\alpha}|T,\xi_{2N},\ldots,\xi_1),T)}{\partial \tilde{\boldsymbol{x}}(s)}\boldsymbol{\varphi}^i(s)$$

$$+\frac{\partial \boldsymbol{g}^{2N+1}(\tilde{\boldsymbol{x}}(s|T,\xi_{2N},\ldots,\xi_1),\tilde{\boldsymbol{x}}(s-\tilde{\alpha}|T,\xi_{2N},\ldots,\xi_1),T)}{\partial \tilde{\boldsymbol{x}}(s-\tilde{\alpha})}\boldsymbol{\varphi}^i(s-\tilde{\alpha}),$$

$$s \in (s_{2N}, 1], \tag{11.51}$$

*with*

$$\boldsymbol{\varphi}^i(s_i) = \boldsymbol{g}^i(\tilde{\boldsymbol{x}}(s_i|T,\xi_{i-1},\ldots,\xi_1),\tilde{\boldsymbol{x}}(s_i-\tilde{\alpha}|T,\xi_{i-1},\ldots,\xi_1),T),$$

$$\boldsymbol{\varphi}^i(s) = \mathbf{0},\ \forall s \in \left[-\bar{\tilde{\alpha}}, s_i\right].$$

*Furthermore,*

$$\frac{\partial \tilde{G}_{\varepsilon,\gamma}(T,\boldsymbol{\xi})}{\partial \xi_{2N+1}} = \sum_{l=1}^{12}\int_0^1 \frac{\partial \varphi_\varepsilon(h_l(\tilde{\boldsymbol{x}}(s|T,\boldsymbol{\xi})))}{\partial h_l}\frac{\partial h_l(\tilde{\boldsymbol{x}}(s|T,\boldsymbol{\xi}))}{\partial \tilde{\boldsymbol{x}}}\boldsymbol{\varphi}^{2N+1}(s)\mathrm{d}s, \tag{11.52}$$

*where*

$$\boldsymbol{\varphi}^{2N+1}(s) = \boldsymbol{g}^{2N+1}(\tilde{\boldsymbol{x}}(s|T,\xi_{2N},\ldots,\xi_1),\tilde{\boldsymbol{x}}(s-\tilde{\alpha}|T,\xi_{2N},\ldots,\xi_1),T),$$

$$s \in (s_{2N}, 1],$$

*with*

$$\boldsymbol{\varphi}^{2N+1}(s) = \mathbf{0},\ s \in \left[-\bar{\tilde{\alpha}}, s_{2N}\right].$$

*Proof.* We can complete the proof using a method similar to the proof of Theorem 11.5. □

In view of the above theorems, we can develop the following computational procedure to generate an approximately optimal solution of (FDOC).

**Algorithm 11.1.**

**Step 1.** Choose initial values of $\varepsilon$, $\gamma$, and $(T,\boldsymbol{\xi})$; set parameters $0 < \beta_1 < 1$, $0 < \beta_2 < 1$, $\bar{\varepsilon}$, and $\bar{\gamma}$.

**Step 2.** Compute $(T^*_{\varepsilon,\gamma}, \boldsymbol{\xi}^*_{\varepsilon,\gamma})$.

**Step 2.1.** Solve the switched time-delay system (11.23)–(11.25) to obtain $\tilde{\boldsymbol{x}}(s|T,\xi_{i-1},\xi_{i-2},\ldots,\xi_1)$, $s \in (s_{i-1}, s_i]$, $i = 1,2,\ldots,2N+1$.

**Step 2.2.** Solve the time-delay systems (11.36)–(11.38) and (11.40)–(11.42) to obtain (11.35), (11.46), (11.39), and (11.50). Furthermore, by (11.44), (11.45), and Step 2.1, we compute (11.43) and (11.52).

**Step 2.3.** Solve ($EOC_{\varepsilon,\gamma}$) using SQP to give $(T^*_{\varepsilon,\gamma}, \boldsymbol{\xi}^*_{\varepsilon,\gamma})$.

**Step 3.** Check feasibility of $G(T^*_{\varepsilon,\gamma}, \boldsymbol{\xi}^*_{\varepsilon,\gamma}) = 0$. If $G(T^*_{\varepsilon,\gamma}, \boldsymbol{\xi}^*_{\varepsilon,\gamma})$ is feasible, then go to Step 4. Otherwise set $\gamma := \beta_1\gamma$. If $\gamma \leqslant \bar{\gamma}$, then we have an abnormal exit. Otherwise go to Step 2.

**Step 4.** Set $\varepsilon := \beta_2\varepsilon$. If $\varepsilon > \bar{\varepsilon}$, then go to Step 2. Otherwise, output $\boldsymbol{\tau}^*_{\varepsilon,\gamma}$ from $(T^*_{\varepsilon,\gamma}, \boldsymbol{\xi}^*_{\varepsilon,\gamma})$ by (11.16) and (11.19) and stop.

Then, $\boldsymbol{\tau}^*_{\varepsilon,\gamma}$ is an approximately optimal solution of (FDOC).

## 11.5 Numerical Results

In the fed-batch fermentation, the reactant composition, cultivation conditions, and determination of biomass, substrate, and metabolites have been reported in [48]. To numerically solve the system (11.1), the initial state, the velocity ratio of adding alkali to substrate, the concentration of initial feed substrate, the feed rate of substrate, the delay argument, and the bound of the delay argument are $\boldsymbol{x}_0 = (0.1115\,\mathrm{g\,L^{-1}}, 495\,\mathrm{mmol\,L^{-1}}, 0, 0, 0, 5\,\mathrm{L})^\top$, $r = 0.75$, $c_{s0} = 10{,}762\,\mathrm{mmol\,L^{-1}}$, $v = 2.25873 \times 10^{-4}\,\mathrm{L\,s^{-1}}$, $\alpha = 0.4652\,\mathrm{h}$, and $\bar{\alpha} = 1\,\mathrm{h}$, respectively. The initial vector of switching instants and the terminal time are taken as the ones in Chap. 6. In addition, the initial function $\tilde{\boldsymbol{\phi}}(t)$ is obtained by interpolating the experimental data with cubic spline method [189].

In order to save computational time, the maximal duration of fed-batch process is partitioned into the first batch phase (Bat. Ph.) and phases I–IX (Phs. I–IX) according to the number of switchings. The same time durations of feed processes (resp. batch processes) are adopted in each one of Phs. I–IX. It should be mentioned that this approach has been adopted to calculate the optimal control in Chap. 6. Moreover, the bounds of the time durations in Bat. Ph. and in each one of Phs. I–IX are as given in Table 7.1.

The delay-differential equations in the computation process are numerically integrated by combination of the fourth-order Runge–Kutta integration scheme and the method of steps with the relative error tolerance $10^{-6}$. All the computations are performed in Visual C++ 6.0 and numerical results are plotted by Matlab 7.10.0 (The Mathworks Inc.) on an AMD Athlon 64 X2 Dual Core Processor TK-57 1.90 GHz machine. Applying Algorithm 11.1 to the (FDOC), we obtain the optimal terminal time $T^* = 17.4609\,\mathrm{h}$, in which the corresponding $N^* = 440$, and the optimal switching instants in Bat. Ph. and Phs. I–IX as listed in Table 11.1. Here, the parameters $\beta_1$ and $\beta_2$ were chosen as 0.1 and 0.01 until the solution obtained is feasible for the original problem. The process was terminated when $\bar{\varepsilon} = 1.0 \times 10^{-8}$ and $\bar{\gamma} = 1.0 \times 10^{-7}$. It is worth mentioning that in the former stage of iterations, a small value of $\gamma$ was required to ensure feasibility. After that the $\gamma$ hardly changed as $\varepsilon$ was decreased.

For the obtained optimal terminal time, it is much shorter than the original terminal time 24.16 h, which is important to reduce the operation costs.

**Table 11.1** The optimal switching instants in fed-batch process.

| Phases | Switching instants | Optimal values (s) |
|---|---|---|
| Bat. Ph. | $\tau_1$ | 18,369.072 |
| Ph. I | $\tau_{2j}$ | $18{,}373.0366 + 100.004(j-1)$ |
| $(j = 1, \ldots, 28)$ | $\tau_{2j+1}$ | $18{,}369.072 + 100.004j$ |
| Ph. II | $\tau_{2j}$ | $21{,}173.191 + 96.003(j-29)$ |
| $(j = 29, \ldots, 65)$ | $\tau_{2j+1}$ | $21{,}169.188 + 96.003(j-28)$ |
| Ph. III | $\tau_{2j}$ | $24{,}728.714 + 103.4065(j-66)$ |
| $(j = 66, \ldots, 126)$ | $\tau_{2j+1}$ | $24{,}721.308 + 103.4065(j-65)$ |
| Ph. IV | $\tau_{2j}$ | $31{,}036.6746 + 101.663(j-127)$ |
| $(j = 127, \ldots, 245)$ | $\tau_{2j+1}$ | $31{,}029.084 + 101.663(j-126)$ |
| Ph. V | $\tau_{2j}$ | $43{,}133.7714 + 99.7768(j-246)$ |
| $(j = 246, \ldots, 378)$ | $\tau_{2j+1}$ | $43{,}126.92 + 99.7768(j-245)$ |
| Ph. VI | $\tau_{2j}$ | $56{,}405.24 + 99.7768(j-379)$ |
| $(j = 379, \ldots, 440)$ | $\tau_{2j+1}$ | $56{,}397.24 + 104.0056(j-378)$ |

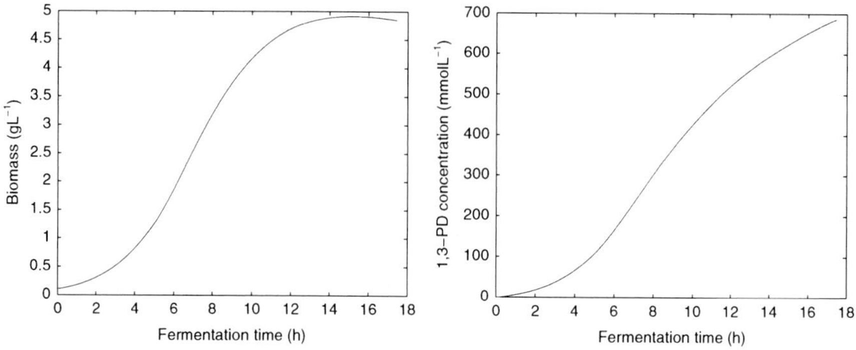


**Fig. 11.1** Concentration profiles of biomass and 1,3-PD in fed-batch process

Moreover, under the obtained optimal switching instants and the optimal terminal time, the maximal mass of 1,3-PD per unit time $J^*$ is 279.591 mmol h$^{-1}$. Under the obtained optimal switching instants and the optimal terminal time, the optimal concentration profiles of biomass and 1,3-PD in the fed-batch process are shown in Fig. 11.1. More importantly, the optimal computed profile of the mass of 1,3-PD per unit time is depicted in Fig. 11.2.

## 11.6 Conclusion

In this chapter, we investigated optimal control of switched time-delay systems in constantly fed-batch process. The free time-delayed optimal control problem was presented. Using the time-scaling transformation and parameterizing the switching

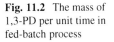

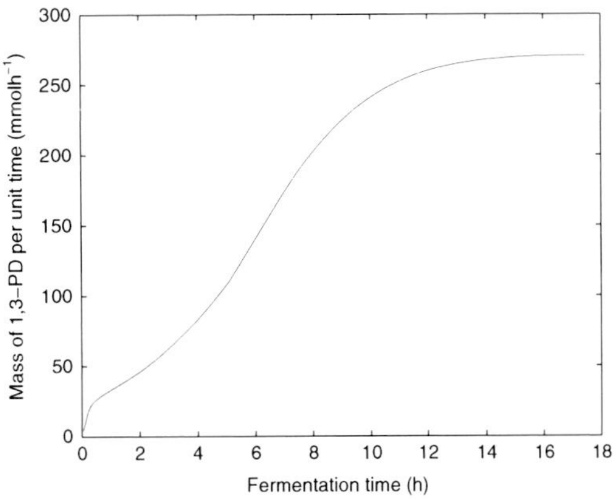


**Fig. 11.2** The mass of 1,3-PD per unit time in fed-batch process

instants into new parameters, the optimal control problem was transcribed into its equivalent form. A computational approach was developed to seek the optimal control strategy. Numerical simulation results verified the effectiveness of the numerical solution method.

# References


1. Ahmed, N.U., Teo, K.L.: Optimal Control of Distributed Parameter Systems. North Holland, New York (1981)
2. an der Heiden, U.: Delays in physiological systems. J. Math. Biol. **8**, 345–364 (1979)
3. Andres-Toro, B., Giron-Sierra, J.M., Lopez-Orozco, J.A., Fernandez-Conde, C.: Application of genetic algorithms and simulations for the optimization of batch fermentation control. Proc. IEEE Int. Conf. Syst. Man Cybern. **1**, 392–397 (1997)
4. Andriantsoa, M., Laget, M., Cremieux, A., Dumenil, G.: Constant fed-batch culture of methanol-utilizing corynebacterium producing vitamin B 12. Biotechnol. Lett. **6**, 783–788 (1984)
5. Arrowsmith, D.K., Place, C.M.: Ordinary Differential Equations. Chapman and Hall, London (1982)
6. Aubin, J.P., Cellina, A.: Differential Inclusions. Springer, Berlin (1984)
7. Augustin, D., Maurer, H.: Second order sufficient conditions and sensitivity analysis for optimal multiprocess control problems. Control Cybern. **29**, 11–31 (2000)
8. Augustin, D., Maurer, H.: Computational sensitivity analysis for state constrained optimal control problems. Ann. Oper. Res. **101**, 75–99 (2001)
9. Axelsson, H., Wardi, Y., Egerstedt, M., Verriest, E.: A gradient descent approach to optimal mode scheduling in hybrid dynamical systems. J. Optim. Theory Appl. **136**, 167–186 (2008)
10. Azema, P., Durante, C., Roubellat, F., Sevely, Y.: Study of the sensitivity of systems to time-delay variations. Electron. Lett. **3**, 171–172 (1967)
11. Babaali, M., Egerstedt, M.: Observability for switched linear systems. In: Rajeev, A., George, J.P. (eds.) Hybrid Systems: Computation and Control. Springer, New York (2004)
12. Babu, B.V., Angira, R.: Modified differential evolution (MDE) for optimization of non-linear chemical processes. Comput. Chem. Eng. **30**, 989–1002 (2006)
13. Bailey, J.E., Ollis, D.F.: Biochemical Engineering Fundamentals. McGraw-Hill, New York (1986)
14. Baker, C.T.H., Bocharov, G.A., Paul, C.A.H., Rihan, F.A.: Modelling and analysis of time-lags in some basic patterns of cell proliferation. J. Math. Biol. **37**, 341–371 (1998)
15. Banks, H.T., Bortz, D.M.: A parameter sensitivity methodology in the context of HIV delay equation models. J. Math. Biol. **50**, 607–625 (2005)
16. Barbirato, F., Himmi, E.H., Conte, T., Bories, A.: 1,3-propanediol production by fermentation: an interesting way to valorize glycerin from the ester and ethanol industries. Ind. Crops Prod. **7**, 281–289 (1998)




C. Liu, Z. Gong, *Optimal Control of Switched Systems Arising in Fermentation Processes*, Springer Optimization and Its Applications 97,
DOI 10.1007/978-3-662-43793-3

17. Barton, P.I., Allgor, R.J., Feehery, W.F., Galan, S.: Dynamic optimization in a discontinuous world. Ind. Eng. Chem. Res. **37**, 966–981 (1998)
18. Bastin, G., Van Impe, J.F.: Nonlinear and adaptive control in biotechnology: a tutorial. Eur. J. Control **1**, 37–53 (1995)
19. Bazaraa, M.S., Sherali, H.D., Shetty, C.M.: Nonlinear Programming Theory and Algorithms. Wiley-Interscience, New York (2006)
20. Bean, J.C., Hadj-Alouane, A.B.: A dual genetic algorithm for bounded integer programs. Technical Report TR 92-53, Department of Industrial and Operations Engineering, The University of Michigan (1992)
21. Bellman, R.E.: Dynamic Programming. Princeton University Press, Princeton (1957)
22. Bemporad, A., Ferrari-Trecate, G., Morari, M.: Observability and controllability of piecewise affine and hybrid systems. IEEE Trans. Autom. Control **45**, 1864–1876 (2000)
23. Bemporad, A., Giua, A., Seatzu, C.: Synthesis of state-feedback optimal controllers for continuous time switched linear systems. In: Proceedings of the 41st IEEE Conference on Decision and Control, Las Vegas, pp. 3182–3187 (2002)
24. Bemporad, A., Giua, A.,Seatzu, C.: A master-slave algorithm for the optimal control of continuous-time switched affine systems. In: Proceedings of the 41st IEEE Conference on Decision and Control, Las Vegas, pp. 1976–1981 (2002)
25. Bemporad, A., Morari, M.: Control of systems integrating logic, dynamics, and constraints. Automatica **35**, 407–427 (1999)
26. Bengea, S.C., Decarlo, R.A.: Optimal and suboptimal control of switching systems. In: Proceedings of the 42nd IEEE Conference on Decision and Control, Maui, pp. 5295–5300 (2003)
27. Bengea, S.C., Decarlo, R.A.: Optimal control of two-switched linear systems. Control Eng. Appl. Inform. **5**, 11–16 (2003)
28. Bengea, S.C., Decarlo, R.A.: Optimal control of switching systems. Automatica **41**, 11–27 (2005)
29. Berkovitz, L.D.: Optimal Control Theory. Springer, New York (1974)
30. Betts, J.T., Gablonsky, J.M.: A comparison of interior point and SQP methods on optimal control problems. Mathematics and computing technology reports M&CT-Tech-02-004, The Boeing Company (2002)
31. Biebl, H., Menzel, K., Zeng, A.P.: Microbial production of 1,3-propanediol. Appl. Microbiol. Biotechnol. **52**, 289–297 (1999)
32. Blom, H.A.P., Bar-Shalom, Y.: The interacting multiple model algorithm for systems with Markovian switching coefficients. IEEE Trans. Autom. Control **33**, 780–783 (2002)
33. Blondel, V.D., Tsitsiklis, J.N.: Complexity of stability and controllability of elementary hybrid systems. Automatica **35**, 479–490 (1999)
34. Bonvin, D.: Optimal operation of batch reactors view a personal. J. Process Control **8**, 355–368 (1998)
35. Branicky, M.: Stability of switched and hybrid systems. In: Proceedings of the 33rd IEEE Conference on Decision and Control, Lake Buena Vista, pp. 3498–3503 (1994)
36. Branicky, M.S., Borkar, V.S., Mitter, S.K.: A unified framework for hybrid control: model and optimal control theory. IEEE Trans. Autom. Control **43**, 31–45 (1998)
37. Breakwell, J.V.: The optimization of trajectories. J. Soc. Ind. Appl. Math. **7**, 215–247 (1959)
38. Bryson, A., Ho, Y.C.: Applied Optimal Control. Halsted Press, New York (1975)
39. Bukovskiy, A.G.: Distributed Control Systems. American Elsevier, New York (1969)
40. Caldwell, T., Murphy, T.: An adjoint method for second-order switching time optimization. In: Proceedings of the 49th IEEE Conference on Decision and Control, Atlanta, pp. 2155–2162 (2010)
41. Cesari, L.: An existence theorem in problems of optimal control. SIAM J. Control **3**, 7–22 (1965)
42. Cesari, L.: Existence theorems for optimal solutions in Pontryagin and Lagrange problems. SIAM J. Control **3**, 475–498 (1966)
43. Cesari, L.: Optimization-Theory and Applications. Springer, New York (1983)

44. Chai, Q., Loxton, R., Teo, K.L., Yang, C.: Time-delay estimation for nonlinear systems with piecewise-constant input. Appl. Math. Comput. **219**, 9543–9560 (2013)
45. Chai, Q., Loxton, R., Teo, K.L., Yang, C.: A class of optimal state-delay control problems. Nonlinear Anal.: Real World Appl. **14**, 1536–1550 (2013)
46. Chai, Q., Loxton, R., Teo, K.L., Yang, C.: A unified parameter identification method for nonlinear time-delay systems. J. Ind. Manag. Optim. **9**, 471–486 (2013)
47. Chan, W.C., Aly, G.M.: A computational method for optimal control problems with a free final time using the modified quasilinearization and the gradient method. Int. J. Control **18**, 1067–1075 (1973)
48. Chen, X., Zhang, D.J., Qi, W.T., Gao, S.J., Xiu, Z.L., Xu, P.: Microbial fed-batch production of 1,3-propanediol by *Klebsiella pneumoniae* under microaerobic conditions. Appl. Microbiol. Biotechnol. **63**, 143–146 (2003)
49. Cheng, K.K., Zhang, J.N., Liu, D.H., Sun, Y., Liu, H.J.: Pilot-scale production of 1,3-propanediol using *Klebsiella pneumoniae*. Process Biochem. **42**, 740–744 (2007)
50. Chiou, J.P., Wang, F.S.: Hybrid method of evolutionary algorithms for static and dynamic optimization problems with application to a fed-batch fermentation process. Comput. Chem. Eng. **23**, 1277–1291 (1999)
51. Clarke, F.H., Ledyaev, S., Stern, R.J.: Nonsmooth Analysis and Control Theory. Springe, New York (1998)
52. Clarke, F.H., Vinter, R.B.: Optimal multiprocesses. SIAM J. Control Optim. **27**, 1072–1091 (1989)
53. Clarke, F.H., Vinter, R.B.: Applications of optimal multiprocesses. SIAM J. Control Optim. **27**, 1048–1071 (1989)
54. Clarke, F.H., Watkins, G.: Necessary conditions, controllability and the value function for differential-difference inclusions. Nonlinear Anal.-Theory Methods Appl. **10**, 1155–1179 (1986)
55. Clarke, F.H., Wolenski, P.R.: The sensitivity of optimal control problems to time delay. SIAM J. Control Optim. **29**, 1176–1215 (1991)
56. Collins, P., van Schuppen, J.H.: Observability of piecewise-affine hybrid systems. In: Hybrid Systems: Computation and Control. Volume 2993 of Lecture Notes in Computer Science, pp. 265–279. Springer, Berlin (2004)
57. Conway, J.B.: A Course in Functional Analysis. Springer, New York (1990)
58. Curtain, R.F., Pritchard, A.J.: Infinite Dimensional Linear Systems Theory. Springer, Berlin (1978)
59. Daniel, L.: Switching in Systems and Control. Birhäuser, Boston (2003)
60. D'Ans, G., Koxotowc, P., Gottlieb, D.: Time-optimal control for a model of bacterial growth. J. Optim. Theory Appl. **7**, 61–69 (1971)
61. D'Ans, G., Koxotowc, P., Gottlieb, D.: A nonlinear regulator problem for a model of biological waste treatment. IEEE Trans. Autom. Control **16**, 341–347 (1971)
62. D'Ans, G., Koxotowc, P., Gottlieb, D.: Optimal control of bacterial growth. Automatica **8**, 729–736 (1972)
63. Dayawansa, W.P., Martin, C.F.: A converse Lyapunov therorem for a class of dynamical systems which undergo switching. IEEE Trans. Autom. Control **44**, 751–760 (1999)
64. DeCarlo, R., Branicky, M., Pettersson, S., Lennartson, B.: Perspectives and results on the stability and stabilizability of hybrid systems. Proc. IEEE **88**, 1069–1082 (2000)
65. Deckwer, W.D.: Microbial conversion of glycerol to 1,3-propanediol. FEMS Microbiol. Rev. **16**, 143–149 (1995)
66. Delmotte, F., Verriest, E.I., Egestedt, M.: Optimal impulsive control of delay systems. ESAIM Control Optim. Calc. Var. **14**, 767–779 (2008)
67. De Schutter, B., Heemels, W.P.M.H., Lunze, J., Prieur, C.: Survey of modeling, analysis, and control of hybrid systems. In: Lunze, J., Lamnabhi-Lagarrigue, F. (eds.) Handbook of Hybrid Systems Control-Theory, Tools, Applications. Cambridge University Press, Cambridge (2009)

68. Dolcetta, J.C., Evans, L.C.: Optimal switching for ordinary differential equations. SIAM J. Control Optim. **22**, 143–161 (1984)
69. Dontchev, A.S., Hager, W.W., Poore, A.B.: Optimality, stability and convergence in nonlinear control. Appl. Math. Optim. **31**, 297–326 (1995)
70. Driver, R.D.: Ordinary and Delay Differential Equations. Springer, New York (1977)
71. Egerstedt, M., Ögren, P., Shakernia, O., Lygeros, J.: Toward optimal control of switched linear systems. In: Proceedings of the 39th IEEE Conference on Decision and Control, Sydney, pp. 587–592 (2000)
72. Egerstedt, M., Wardi, Y., Delmotte, F.: Optimal control of switching times in switched dynamical systems. In: Proceedings of the 42nd IEEE Conference on Decision and Control, Maui, pp. 2138–2143 (2003)
73. Elnagar, G.N., Kazemi, M.A.: Pseudospectral Chebyshev optimal control of constrained nonlinear dynamical systems. Comput. Optim. Appl. **11**, 195–217 (1998)
74. Elnagar, G.N., Kazemi, M.A., Razzaghi, M.: The pseudospectral Legendre method for discretizing optimal control problems. IEEE Trans. Autom. Control **40**, 1793–1796 (1995)
75. Eslami, M.: Theory of Sensitivity in Dynamic Systems. Springer, New York (1994)
76. Ezzine, J., Haddad, A.H.: Controllability and observability of hybrid systems. Int. J. Control **49**, 2045–2055 (1989)
77. Fahroo, F., Ross, I.M.: Direct trajectory optimization pseudospectral method. J. Guid. Control Dyn. **25**, 160–166 (2002)
78. Filippov, A.F.: On certain questions in the theory of optimal control. SIAM J. Control Optim. **1**, 76–84 (1962)
79. Fleming, W.H., Rishel, R.W.: Deterministic and Stochastic Optimal Control. Springer, Berlin (1975)
80. Fletcher, R.: Practical Methods of Optimization. Volume 2 Constrained Optimization. Wiley, New York (1981)
81. Folland, G.B.: Real Analysis. Wiley, New York (1999)
82. Forage, R., Lin, E.C.C.: *dha* system mediating aerobic and anaerobic dissimilation of glycerol in *Klebsiella pneumoniae* NCIB418. J. Bacteriol. **15**, 591–599 (1982)
83. Freund, A.: Uber die Bildung und Darstellung von Trimethylenalkohol aus Glycerin. Monatsh. Chimie **2**, 636–641 (1881)
84. Gao, C.X., Feng, E.M., Wang, Z.T., Xiu, Z.L.: Nonlinear dynamical systems of bio-dissimilation of glycerol to 1,3-propanediol and their optimal controls. J. Ind. Manag. Optim. **1**, 377–388 (2005)
85. Gao, J.G., Shen, B.Y., Feng, E.M., Xiu, Z.L.: Modelling and optimal control for an impulsive dynamical system in microbial fed-batch culture. Comput. Appl. Math. **32**, 275–290 (2013)
86. Ge, S.S., Sun, Z.D., Lee, T.H.: Reachability and controllability of switched linear discrete-time systems. IEEE Trans. Autom. Control **46**, 1437–1441 (2001)
87. Giua, A., Seatzu, C., Van der Mee, C.M.: Optimal control of autonomous linear systems switched with a preassigned finite sequence. In: Proceedings of the 2001 IEEE International Symposium on Intelligent Control, Mexico City, pp. 144–149 (2001)
88. Göllmann, L., Kern, D., Maurer, H.: Optimal control problems with delays in state and control variables subject to mixed control-state constraints. Optimal Control Appl. Methods **30**, 341–365 (2009)
89. Goncalves, J.M., Megretski, A., Dahleh, M.A.: Global analysis of piecewise linear systems using impact maps and surface Lyapunov functions. IEEE Trans. Autom. Control **48**, 2089–2106 (2003)
90. Gong, Z.H.: A multistage system of microbial fed-batch fermentation and its parameter identification. Math. Comput. Simul. **80**, 1903–1910 (2010)
91. Gong, Z.H., Liu, C.Y., Feng, E.M., Wang, L., Yu, Y.S.: Modelling and optimization for a switched system in microbial fed-batch culture. Appl. Math. Model. **35**, 3276–3284 (2011)
92. Gorbunov, V.K., Lutoshkin, I.V.: Development and experience of using the parameterization method in singular problems of dynamic optimization. J. Comput. Syst. Sci. Int. **43**, 725–742 (2004)

93. Gurramkonda, C., et al.: Application of simple fed-batch technique to high-level secretory production of insulin precursor using *Pichia pastoris* with subsequent purification and conversion to human insulin. Microb. Cell Factories **9**, 1–11 (2010)
94. Hairer, E., Nørsett, S.P., Wanner, G.: Solving Ordinary Differential Equations I: Nonstiff Problems. Springer, Berlin (2009)
95. Hale, J.K., Verduyn Lunel, S.M.: Introduction to Functional Differential Equations. Springer, Berlin (1993)
96. Han, S.P.: Superlinearity convergent variable metric algorithm for general nonlinear programming problems. Math. Program. **11**, 263–282 (1976)
97. Han, S.P.: A globally convergent method for nonlinear programming. J. Optim. Theory Appl. **22**, 297–309 (1977)
98. Hayashi, Y., Matsuki, J., Kanai, G.: Application of improved PSO to power flow control by TCSC for maximum acceptance of requested wheeled power. Translated from Denki Gakkai Ronbunshi **10**, 1133–1141 (2003)
99. Heil, C.: A Basic Theory Primer. Springer, New York (1998)
100. Hespanha, J., Liberzon, D., Morse, A.S.: Overcoming the limitations of adaptive control by means of logic-based switching. Syst. Control Lett. **49**, 49–56 (2003)
101. Hicks, G.H., Ray, W.H.: Approximation methods for optimal control synthesis. Can. J. Chem. Eng. **49**, 522–528 (1971)
102. Hirschmann, S., Baganz, K., Koschik, I., Vorlop, K.D.: Development of an integrated bioconversion process for the production of 1,3-propanediol from raw glycerol waters. Landbauforsch. Völkenrode **55**, 261–267 (2005)
103. Hjersted, J., Henson, M.A.: Population modeling for ethanol productivity optimization in fed-batch yeast fermenters. In: Proceedings of American Control Conference, Portland, pp. 3253–3258 (2005)
104. Holland, J.H.: Adaptaion in Natural and Artificial Systems. The University of Michigan Press, Michigan (1975)
105. Homaifar, A., Lai, S.H.Y., Qi, X.: Constrained optimization via genetic algorithms. Simulation **62**, 242–254 (1994)
106. Hou, L., Michel, A.N., Ye, H.: Stability analysis of switched systems. In: Proceedings of the 35th IEEE Conference on Decision and Control, Kobe, pp. 1208–1212 (1996)
107. Jaddu, H., Shimemura, E.: Computational methods based on the state parameterization for solving constrained optimal control problems. Int. J. Syst. Sci. **30**, 275–282 (1999)
108. Jadot, F., Bastin, G., Van Impe, J.F.: Optimal adaptive control of a bioprocess with yield-productivity conflict. J. Biotechnol. **65**, 61–68 (1998)
109. Jennings, L.S., Teo, K.L., Goh, C.J.: MISER3.3, optimal control software: theory and user manual. Department of Mathematics, The University of Western Australia (2000)
110. Jiménez-Hornero, J.E., Santos-Dueñas, I.M., García-García, I.: Optimization of biotechnological processes. The acetic acid fermentation. Part III: dynamic optimization. Biochem. Eng. J. **45**, 22–29 (2009)
111. Johnson, A.: The control of fed-batch fermentation processes-a survey. Automatica **23**, 691–705 (1987)
112. Joines, J.A., Houck, C.R.: On the use of non-stationary penalty functions to solve nonlinear constrained optimization problems with GA's. In: Proceedings of the First IEEE International Conference on Evolutionary Computation, Orlando, pp. 579–584 (1994)
113. Kalman, R.E.: Contribution to the theory of optimal control. Bol. Soc. Mat. Mex., **5**, 102–119 (1960)
114. Kamien, M.I., Schwartz, N.L.: Dynamic Optimization-The Calculus of Variations and Optimal Control in Economics and Management. Elsevier Sciences B.V., Amsterdam (1991)
115. Kaya, C.Y., Noakes, J.L.: Computational method for time-optimal switching control. J. Optim. Theory Appl. **117**, 69–92 (2003)
116. Keller, H.B.: Numerical Methods for Two-Point Boundary Value Problems. Dover, New York (1992)

117. Kennedy, J., Eberhart, R.C.: Particle swarm optimization. In: Proceedings of the 1995 IEEE International Conference on Neural Networks, Perth, pp. 1942–1948 (1995)
118. Kennedy, J., Spears, W.M.: Matching algorithms to problems: an experimental test of the particle swarm and some genetic algorithms on the multimodal problem generator. In: Proceedings of IEEE International Conference on Evolutionary Computation, Anchorage, pp. 74–77 (1998)
119. Kharatishvili, G.L.: Maximum principle in the theory of optimal time-delay processes. Dokl. Akad. Nauk SSSR **136**, 39–43 (1961)
120. Kharatishvili, G.L.: A Maximum Principle in External Problems with Delays, Mathematical Theory on Control. Academic, New York (1967)
121. Kim, D.K., Park, P.G., Ko, J.W.: Output-feedback $H_\infty$ control of systems over communication networks using a deterministic switching system approach. Automatica **40**, 1205–1212 (2004)
122. Kirkpatrick, S., Gelatt, C.U., Vechhi, M.P.: Optimization by simulated annealing. Science **220**, 671–680 (1983)
123. Kleban, J.: Switched Systems. In-Teh, Vukovar (2009)
124. Koda, M.: Sensitivity analysis of time-delay systems. Int. J. Syst. Sci. **12**, 1389–1397 (1981)
125. Korytowski, A., Szymkat, M., Maurer, H., Vossen, G.: Optimal control of a fedbatch fermentation process: numerical methods, sufficient conditions and sensitivity analysis. In: Proceedings of the 47th IEEE Conference on Decision and Control, Cancún, pp. 1551–1556 (2008)
126. Koziel, S., Michalewicz, Z.: Evolutionary algorithms, homomorphous mappings, and constrained parameter optimization. Evol. Comput. **7**, 19–44 (1999)
127. Kraft, D.: On converting optimal control problems into nonlinear programming problems. In: Schittkowski, K. (ed.) Computational Mathematical Programming. Springer, Berlin, 261–280 (1985)
128. Kulkarniand, S.R., Ramadge, P.J.: Model and controller selection policies based on output prediction errors. IEEE Trans. Autom. Control **41**, 1594–1604 (1996)
129. Lee, J., Lee, S., Park, S., Middeelberg, A.: Control of fed-batch fermentations. Biotechnol. Adv. **17**, 29–48 (1999)
130. Lee, K.K., Arapostathis, A.: On the controllability of piecewise linear hypersurface systems. Syst. Control Lett. **9**, 89–96 (1987)
131. Lee, S.Y., Hong, S.H., Lee, S.H., Park, S.J.: Fermentative production of chemicals that can be used for polymer synthesis. Macromol. Biosci. **4**, 157–164 (2004)
132. Li, H.Q., Li, L., Kim, T.H., Xie, S.L.: An improved PSO-based of harmony search for complicated optimization problems. Int. J. Hybrid Inf. Technol. **1**, 57–64 (2008)
133. Li, R., Teo, K.L., Wong, K.H., Duan, G.R.: Control parameterization enhancing transform for optimal control of switched systems. Math. Comput. Model. **43**, 1393–1403 (2006)
134. Liberzon, D., Morse, A.S.: Basic problems in stability and design of switched systems. IEEE Control Syst. Mag. **19**, 59–70 (1999)
135. Lim, H.C., Chen, B.J., Creagan, C.C.: An analysis of extended and exponentially-fed-batch cultures. Biotechnol. Bioeng. **1**, 425–429 (1977)
136. Lim, H.C., Tayeb, Y.J., Modak, J.M., Bonte, P.: Computational algorithms for optimal feed rates for a class of fed-batch fermentation: numerical results for penicillin and cell mass production. Biotechnol. Bioeng. **28**, 1408–1420 (1986)
137. Lin, H., Antsaklis, P.J.: Stability and stabilizability of switched linear systems: a survey of recent results. IEEE Trans. Autom. Control **54**, 308–322 (2009)
138. Lin, Q., Loxton, R., Teo, K.L.: Optimal control of nonlinear switched systems: computational methods and applications. J. Oper. Res. Soc. China, **1**, 275–311 (2013)
139. Lin, Q., Loxton, R., Teo, K.L.: The control parameterization method for nonlinear optimal control: a survey. J. Ind. Manag. Optim. **10**, 275–309 (2014)
140. Lin, Q., Loxton, R., Teo, K.L., Wu, Y.H.: A new computational method for a class of free terminal time optimal control problems. Pac. J. Optim. **7**, 63–81 (2011)

141. Lin, Q., Loxton, R., Teo, K.L., Wu, Y.H.: Optimal control computation for nonlinear systems with state-dependent stopping criteria. Automatica **48**, 2116–2129 (2012)
142. Lions, J.L.: Optimal Control of Systems Governed by Partial Differential Equations. Springer, Berlin (1971)
143. Liu, C.Y.: Optimal control for nonlinear dynamical system of microbial fed-batch culture. J. Comput. Appl. Math. **232**, 252–261 (2009)
144. Liu, C.Y.: Modelling and parameter identification for a nonlinear time-delay system in microbial batch fermentation. Appl. Math. Model. **37**, 6899–6908 (2013)
145. Liu, C.Y.: Sensitivity analysis and parameter identification for a nonlinear time-delay system in microbial fed-batch process. Appl. Math. Model. **38**, 1449–1463 (2014)
146. Liu, C.Y., Feng, E.M.: Optimal control of switched autonomous systems in microbial fed-batch cultures. Int. J. Comput. Math. **88**, 396–407 (2011)
147. Liu, C.Y., Gong, Z.H.: Modelling and optimal control of a time-delayed switched system in fed-batch process. J. Frankl. Inst. **351**, 840–856 (2014)
148. Liu, C.Y., Gong, Z.H., Feng, E.M.: Optimal control for a nonlinear time-delay system in fed-batch fermentation. Pac. J. Optim. **9**, 595–612 (2013)
149. Liu, C.Y., Gong, Z.H., Feng, E.M., Yin, H.C.: Optimal switching control for microbial fed-batch culture. Nonlinear Anal.: Hybrid Syst. **2**, 1168–1174 (2008)
150. Liu, C.Y., Gong, Z.H., Feng, E.M., Yin, H.C.: Modelling and optimal control for nonlinear multistage dynamical system of microbial fed-batch culture. J. Ind. Manag. Optim. **5**, 835–850 (2009)
151. Liu, C.Y., Gong, Z.H., Feng, E.M., Yin, H.C.: Optimal switching control of a fed-batch fermentation process. J. Glob. Optim. **52**, 265–280 (2012)
152. Liu, C.Y., Gong, Z.H., Shen, B.Y., Feng, E.M.: Modelling and optimal control for a fed-batch fermentation process. Appl. Math. Model. **37**, 695–706 (2013)
153. Loeblein, C., Perkins, J.D., Srinivasan, B., Bonvin, D.: Economic performance analysis in the design of on-line batch optimization systems. J. Process Control **9**, 61–78 (1999)
154. Loewen, P.D., Rockafellar, R.T.: New necessary conditions for the generalized problem of Bolza. SIAM J. Control Optim. **34**, 1496–1551 (1996)
155. Loewen, P.D., Rockafellar, R.T.: Bolza problems with general time constraints. SIAM J. Control Optim. **35**, 2050–2069 (1997)
156. Loxton, R., Lin, Q., Rehbock, V., Teo, K.L.: Control parameterization for optimal control problems with continuous inequality constraints: new convergence results. Numer. Algebra Control Optim. **2**, 571–599 (2012)
157. Loxton, R.C., Teo, K.L., Rehbock, V.: Optimal control problems with multiple characteristic time points in the objective and constraints. Automatica **44**, 2923–2929 (2008)
158. Loxton, R.C., Teo, K.L., Rehbock, V.: Computational method for a class of switched system optimal control problems. IEEE Trans. Autom. Control **54**, 2455–2460 (2009)
159. Loxton, R.C., Teo, K.L., Rehbock, V.: An optimization approach to state-delay identification. IEEE Trans. Autom. Control **55**, 2113–2119 (2010)
160. Loxton, R.C., Teo, K.L., Rehbock, V., Ling, W.K.: Optimal switching instants for a switched capacitor DC/DC power converter. Automatica **45**, 973–980 (2009)
161. Loxton, R.C., Teo, K.L., Rehbock, V., Yiu, K.F.C.: Optimal control problems with a continous inequality constraint on the state and the control. Automatica **45**, 2250–2257 (2009)
162. Luenberger, D.G., Ye, Y.Y.: Linear and Nonlinear Programming. Springer, New York (2008)
163. Luus, R.: Piecewise linear continuous optimal control by iterative dynamic programing. Ind. Eng. Chem. Res. **32**, 856–865 (1993)
164. MacDonald, N.: Time-Lags in Biological Models. Lecture Notes in Biomathematics, vol. 27. Springer, Berlin (1979)
165. Mahmoud, M.S.: Switched Time-delay Systems: Stablity and Control. Springer, New York (2010)
166. Malanowski, K., Maurer, H.: Sensitivity analysis for parametric control problems with control-state constraints. Comput. Optim. Appl. **5**, 253–283 (1996)

167. Matsui, T., Kato, K., Sakawa, M., Uno, T., Morihara, K.: Nonlinear programming based on particle swarm optimization. In: Chan, A.H.S., Ao, S.I. (eds.) Advances in Industrial Engineering and Operation Research, pp. 173–183. Springer, New York (2008)
168. Maurer, H., Oberle, H.J.: Second order sufficient conditions for optimal control problems with free final time: the Riccati approach. SIAM J. Control Optim. **41**, 380–403 (2002)
169. Maurer, H., Pesch, J.: Solution differentiability for parametric nonlinear control problems with control-state constraints. Control Cybern. **23**, 201–227 (1994)
170. McNeil, B., Harvey, L.M.: Practical Fermentation Technology. Wiley, Chichester (2008)
171. Mendes, P., Kell, D.B.: Nonlinear optimization of biochemical pathways: applications to metabolic engineering and parameter estimation. Bioinformatics **14**, 869–883 (1998)
172. Menzel, K., Zeng, A.P., Biebl, H., Deckwer, W.D.: Kinetic, dynamic, and pathway studies of glycerol metabolism by *Klebsiella pneumoniae* in anaerobic continuous culture: I. The phenomena and characterization of oscillation and hysteresis. Biotechnol. Bioeng. **52**, 549–560 (1996)
173. Menzel, K., Zeng, A.P., Deckwer, W.D.: High concentration and productivity of 1,3-propanediol from continuous fermentation of glycerol by *Klebsiella pneumoniae*. Enzyme Microb. Technol. **20**, 82–86 (1997)
174. Mereau, P.M., Powers, W.F.: A direct sufficient condition for free final time optimal control problems. SIAM J. Control Optim. **14**, 613–622 (1976)
175. Meyer, C., Schroder, S., De Doncker, R.W.: Solid-state circuit breakers and current limiters for medium-voltage systems having distributed power systems. IEEE Trans. Power Electron. **19**, 1333–1340 (2004)
176. Mhaskar, P., El-Farra, N.H., Christofides, P.D.: Predictive control of switched nonlinear systems with scheduled mode transitions. IEEE Trans. Autom. Control **50**, 1670–1680 (2005)
177. Miele, A.: Method of particular solutions for linear two-point boundary-value problems. J. Optim. Theory Appl. **2**, 315–334 (1968)
178. Mignone, D., Ferrari-Trecate, G., Morari, M.: Stability and stabilization of piecewise affine and hybrid systems: an LMI approach. In: Proceedings of the 39th IEEE Conference on Decision and Control, Sydney, pp. 504–509 (2000)
179. Modak, J.M., Lim, H.C., Tayeb, Y.J.: General characteristics of optimal feed rate profiles for various fed-batch fermentation processes. Biotechnol. Bioeng. **28**, 1396–1407 (1986)
180. Moles, C.G., Mendes, P., Banga, J.R.: Parameter estimation in biochemical pathways: a comparison of global optimization methods. Genome Res. **13**, 2467–2474 (2003)
181. Morse, A.S.: Supervisory control of families of linear set-point controllers, part I: exact matcthing. IEEE Trans. Autom. Control **41**, 1411–1431 (1996)
182. Müller, M.A., Martius, P., Allgöwer, F.: Model predictive control of switched nonlinear systems under average dwell-time. J. Process Control **22**, 1702–1710 (2012)
183. Nagy, Z.K., Braatz, R.D.: Open-loop and closed-loop robust optimal control of batch processes using distributional and worst-case analysis. J. Process Control **14**, 411–422 (2004)
184. Nakamura, C.E., et al.: Method for the production of 1,3-propanediol by recombinant microorganisms. US Patent No. 6,013,494 (2000)
185. Nakamura, C.E., Whited, G.M.: Metabolic engineering for the microbial production of 1,3-propanediol. Curr. Opin. Biotechnol. **14**, 454–459 (2003)
186. Narendra, K.S., Balakrishnan, J.: Improving transient response of adaptive control systems using multiple models and switching. IEEE Trans. Autom. Control **39**, 1861–1866 (1994)
187. Narendra, K.S., Balakrishnan, J.: A common Lyapunov function for stable LTI systems with commuting a-matrices. IEEE Trans. Autom. Control **39**, 2469–2471 (1994)
188. Nielsen, J., Villadsen, J.: Modelling of microbial kinetics. Chem. Eng. Sci. **47**, 4225–4270 (1992)
189. Nocedal, J., Wright, S.J.: Numerical Optimization. Springer, New York (1999)
190. Oberle, H.J., Grimm, W.: BNDSCO-a program for the numerical solution of optimal control problems. Institute for Flight Systems Dynamics, DLR, Oberpfaffenhofen, Germany, Internal Report 515-89/22 (1989)

191. Oberle, H.J., Sothmann, B.: Numerical computation of optimal feed rates for a fed-batch fermentation model. J. Optim. Theory Appl. **100**, 1–13 (1993)
192. Oğuztöreli, M.N.: Time Lag Control Systems. Academic, New York (1966)
193. Ohno, H., Nakanishi, E., Takamatsu, T.: Optimal control of a semibatch fermentation. Biotechnol. Bioeng. **18**, 837–864 (1976)
194. O'Sullivan, F.: Sensitivity analysis for regularized estimation in some system identification problems. SIAM J. Sci. Stat. Comput. **12**, 1266–1283 (1991)
195. Panda, B.P., Ali, M., Javed, S.: Fermentation process optimization. Res. J. Microbiol. **2**, 201–208 (2007)
196. Panpanikolaou, S.: Microbial conversion of glycerol into 1,3-propanediol: glycerol assimilation, biochemical events related with 1,3-propanediol biosynthesis and biochemical engineering of the process. In: Aggelis, G.(ed.) Microbial Conversions of Raw Glycerol, pp. 137–168. Nova Science Publishers, New York (2009)
197. Papanikolaou, S., Fick, M., Aggelis, G.: The effect of raw glycerol concentration on the production of 1,3-propanediol by *Clostridium butyricum*. J. Chem. Technol. Biotechnol. **79**, 1189–1196 (2004)
198. Parsopoulos, K.E., Varahatis, M.N.: Recent approaches to global optimization problems through particle swarm optimization. Nat. Comput. **1**, 235–306 (2002)
199. Piccoli, B.: Necessary conditions for hybrid optimization. In: Proceedings of the 38th IEEE Conference on Decision and Control, Phoenix, pp. 410–415 (1999)
200. Polak, E.: Computation Methods in Optimization. Academic, New York (1971)
201. Polak, E.: Optimazation Algorithms and Consistent Approximation. Springer, New York (1997)
202. Pontryagin, L.S., Boltyanski, V.G., Gamkrelidze, V., Mischenko, E.E: The Mathematical Theory of Optimal Control Process. Wiley, New York (1962)
203. Powell, M.J.D.: A fast algorithm for nonlinearly constrained optimization calculations. In: Matson, G.A. (ed.) Numerical Analysis. Lecture Notes in Mathematics, vol. 630. Springer, Berlin (1978)
204. Price, K., Storn, R., Lampinen, J.: Differential Evolution: A Practical Approach to Global Optimization. Springer, Heidelberg (2005)
205. Ramirez, W.F.: Application of Optimal Control Theory to Enhanced Oil Recovery. Elsevier Sciences B.V., Amsterdam (1987)
206. Rani, K.Y., Rao, V.S.R.: Control of fermenters: A review. Bioprocess Eng. **21**, 77–89 (1999)
207. Reimann, A., Biebl, H., Deckwer, W.D.: Production of 1,3-propanediol by *Clostridium butyricum* in continuous culture with cell recycling. Appl. Microbiol. Biotechnol. **49**, 359–363 (1998)
208. Richard, J.P.: Time-delay systems: an overview of some recent advances and open problems. Automatica **29**, 1667–1694 (2003)
209. Rihan, F.A.: Sensitivity analysis for dynamic systems with time-lags. J. Comput. Appl. Math. **151**, 445–462 (2003)
210. Rocha, M., Neves, J., Rocha, I., Ferreira, E.C.: Evolutionary algorithms for optimal control in fed-Batch fermentation processes. In: Gü R.R., et al. (eds.) Applications of Evolutionary Computing, pp. 84–93. Springer, Berlin (2004)
211. Rosenbrock, H., Storey, C.: Computational Techniques for Chemical Engineers. Pergamon Press, Oxford (1966)
212. Ross, I.M., Fahroo, F.: Pseudospectral knotting methods for solving optimal control problems. J. Guid. Control Dyn. **27**, 397–405 (2004)
213. Roubos, J.A., van Straten, G., van Boxtel, A.J.: An evolutionary strategy for fed-batch bioreactor optimization: concepts and performance. J. Biotechnol. **67**, 173–187 (1999)
214. Roxin, E.: The existence of optimal controls. Mich. Math. J. **9**, 109–119 (1962)
215. Rudin, W.: Functional Analysis. McGraw-Hill, New York (1991)
216. Sargent, R.W.H.: Optimal control. J. Comput. Appl. Math. **124**, 361–371 (2000)
217. Sarkar, D., Modak, J.M., Optimization of fed-batch bioreactors using genetic algorithm: multiple control variables. Comput. Chem. Eng. **28**, 789–798 (2004)

218. Sarker, R., Mohammadian, M., Yao, X. (eds.): Evolutionary Optimization. Kluwer Academic, New York (2003)
219. Schwefel, H.P.: Numerical Optimization of Computer Models. Wiley, New York (1981)
220. Seatzu, C., Corona, D., Giua, A., Bemporad, A.: Optimal control of continuous time switched affine systems. IEEE Trans. Autom. Control **51**, 726–741 (2006)
221. Seidman, T.I.: Optimal control of switching systems. In: Proceedings of the 21st Annual Conference on Information Science and Systems, Baltimore, pp. 485–489 (1987)
222. Seierstad, A.: Sufficent conditions in free final time optimal control problems. SIAM J. Control Optim. **26**, 155–167 (1988)
223. Shaikh, M.S., Caines, P.E.: On trajectory optimization for hybrid systems: Theory and algorithms for fixed schedules. In: Proceedings of the 41st IEEE Conference on Decision and Control, Las Vegas, pp. 1997–1998 (2002)
224. Shaikh, M.S., Caines, P.E.: Optimality zone algorithms for hybrid systems computation and control: from exponential to linear complexity. In: Proceedings of the 44th IEEE Conference on Decision and Control/European Control Conference, Seville, pp. 1403–1408 (2005)
225. Shampine, L.F., Thompson, S.: Solving DDEs in Matlab. Appl. Numer. Math. **37**, 441–458 (2001)
226. Shen, B.Y., Liu, C.Y., Ye, J.X., Feng, E.M., Xiu, Z.L.: Parameter identification and optimization algorithm in microbial continuous culture. Appl. Math. Model. **36**, 585–595 (2012)
227. Shen, L.J., Feng, E.M., Wu, Q.D.: Impulsive control in microorganisms continuous fermentation. Int. J. Biomath. **5**, 1250013, 9p (2012)
228. Shi, Y.H., Eberhart, R.C.: A modified particle swarm optimizer. In: Proceedings of IEEE International Conference on Evolutionary Computation, Anchorage, pp. 69–73 (1998)
229. Sienz, J., Innocente, M.S.: Particle swarm optimization: fundamental study and its application to optimization and to jetty scheduling problems. In: Topping, B.H.V., Papadrakakis, M. (eds.) Trends in Engineering Computational Technology, pp. 103–126. Saxe-Coburg Publications, Stirlingshire (2008)
230. Sirisena, H.R., Chou, F.S.: State parameterization approach to the solution of optimal control problems. Optim. Control Appl. Methods **2**, 289–298 (1981)
231. Smets, I.Y.M., Versyck, K.J.E., Van Impe, J.F.M.: Optimal control theory: a generic tool for identification and control of (bio-)chemical reactors. Annu. Rev. Control **26**, 57–73 (2002)
232. Sontag, E.D.: Mathematical Control Theory: Deterministic Finite Dimensional Systems. Springer, New York (1998)
233. Stoddart, A.W.J.: Existence of optimal controls. Pac. J. Math. **1**, 167–177 (1967)
234. Stoer, J., Bulirsch, R.: Introduction to Numerical Analysis. Springer, New York (1980)
235. Storn, R., Price, K.: Differential evolution-a simple and efficient heuristic for global optimization over continuous Spaces. J. Glob. Optim. **11**, 341–359 (1997)
236. Subchan, S., Żbikowski, R.: Computational Optimal Control Tools and Practice. Wiley, Chichester (2009)
237. Sun, Y., Qi, W., Teng, H., Xiu, Z., Zeng, A.: Mathematical modeling of glycerol fermentation by *Klebsiella pneumoniae*: concerning enzyme-catalytic reductive pathway and transport of glycerol and 1, 3-propanediol across cell membrane. Biochem. Eng. J. **38**, 22–32 (2008)
238. Sussmann, H.J.: A maximum principle for hybrid optimal control problems. In: Proceedings of the 38th IEEE Conference on Decision and Control, Phoenix, pp. 425–430 (1999)
239. Takamatsu, T., Hashimoto, I., Shoiya, S., Mizuhara, K., Koike, T., Ohno, H.: Theory and practice of optimal control in continuous fermentation process. Automatica **11**, 141–148 (1975)
240. Teo, K.L., Goh, C.J., Wong, K.H.: A Unified Computational Approach to Optimal Control Problems. Longman Scientific and Technical, Essex (1991)
241. Teo, K.L., Jennings, L.S.: Optimal control with a cost on changing control. J. Optim. Theory Appl. **68**, 335–357 (1991)
242. Teo, K.L., Jennings, L.S., Lee, H.W.J., Rehbock, V.: The control parameterization enhancing transform for constrained optimal control problems. J. Aust. Math. Soc. Ser. B **40**, 314–335 (1999)

243. Teo, K.L., Rehbock, V., Jennings, L.S.: A new computational algorithm for functional inequality constrained optimization problems. Automatica **29**, 789–792 (1993)
244. Teo, K.L., Wu, Z.S.: Computational Methods for Optimizing Distributed Systems. Academic, Orlando (1984)
245. Terwiesch, P., Agarwal, M., Rippin, D.W.T.: Batch unit optimization with imperfect modelling: a survey. J. Process Control **4**, 238–258 (1994)
246. Upreti, S.R.: A new robust technique for optimal control of chemical engineering processes. Comput. Chem. Eng. **28**, 1325–1336 (2004)
247. Van Impe, J.F., Bastin, G.: Optimal adaptive control of fed-batch fermentation processes. Control Eng. Pract. **3**, 939–954 (1995)
248. Verriest, E.I.: Optimal control for switched point delay systems with refractory period. In: Proceedings of the 16th IFAC World Congress, Prague, July 2005
249. Verriest, E.I., Delmotte, F., Egerstedt, M.: Optimal impulsive control of point delay systems with refractory period. In: Proceedings of the 5th IFAC Workshop on Time Delay Systems, Leuven, Sept 2004
250. von Stryk, O., Bulirsch, R.: Direct and indirect methods for trajectory optimization. Ann. Oper. Res. **37**, 357–373 (1992)
251. Wang, F.S., Shyu, C.H.: Optimal feed policy for fed-batch fermentation of ethanol production by *Zymomous mobilis*. Bioprocess Eng. **17**, 63–68 (1997)
252. Wang, G., Feng, E.M., Xiu, Z.L.: Vector measure for explicit nonlinear impulsive system of glycerol bioconversion in fed-batch cultures and its parameter identification. Appl. Math. Comput. **188**, 1151–1160 (2007)
253. Wang, G., Feng, E.M., Xiu, Z.L.: Modelling and parameter identification of microbial biconversion in fed-batch cultures. J. Process Control **18**, 458–464 (2008)
254. Wang, H.Y., Feng, E.M., Xiu, Z.L.: Optimality condition of the nonlinear impulsive system in fed-batch fermentation. Nonlinear Anal.: Theory Methods Appl. **68**, 12–23 (2008)
255. Wang, J., Ye, J.X., Yin, H.C, Feng, E.M., Wang, L.: Sensitivity analysis and identification of kinetic parameters in batch fermentation of glycerol. J. Comput. Appl. Math. **236**, 2268–2276 (2012)
256. Wang, L., Xiu, Z.L., Gong, Z.H., Feng, E.M.: Modeling and parameter identification for multistage simulation of microbial bioconversion in batch culture. Int. J. Biomath. **5**, 1250034, 12p (2012)
257. Wang, L.Y., Gui, W.H., Teo, K.L., Loxton, R., Yang, C.H.: Time delayed optimal control problems with multiple characteristic time points: computation and industrial applications. J. Ind. Manag. Optim. **5**, 705–718 (2009)
258. Wardi, Y., Egerstedt, M.: Algorithm for optimal mode scheduling in switched systems. In: Proceedings of American Control Conference, Montreal, pp. 4546–4551 (2012)
259. Wei, S., Uthaichana, K., Zefran, M., DeCarlo, R.A., Bengea, S.: Applications of numerical optimal control to nonlinear hybrid systems. Nonlinear Anal.: Hybrid Syst. **1**, 264–279 (2007)
260. Wheeden, R.L., Zygmund, A.: Measure and Integral. Marcel Dekker, New York (1977)
261. Wicks, M., DeCarlo, R.: Solution of coupled Lyapunov equations for the stabilization of multimodal linear systems. In: Proceedings of the American Control Conference, New Mexico, pp. 1709–1713 (1997)
262. Wicks, M.A., Pelelies, P., DeCarlo, R.A.: Switched controller synthesis for the quadratic stabilisation of a pair of unstable linear systems. Eur. J. Control **4**, 140–147 (1998)
263. Wilson, R.B.: A simplicial method for convex programming. Ph.D. thesis, Harvard University, Cambridge (1963)
264. Witt, U., Muller, R.J., Augusta, J., Widdecke, H., Deckwer, W.D.: Synthesis, properties and biodegradability of polyesters based on 1,3-propanediol. Macromol. Chem. Phys. **195**, 793–802 (1994)
265. Wong, K.H., Clements, D.J., Teo, K.L.: Optimal control computation for nonlinear time-lag systems. J. Optim. Theory Appl. **47**, 91–107 (1985)
266. Wong, K.H., Jennings, L.S., Benyahz, F.: The control parameterization enhancing transform for constrained time-delayed optimal control problems. ANZIAM J. **43**, 154–185 (2002)

267. Wu, C.Z., Teo, K.L.: Global impulsive optimal control computation. J. Ind. Manag. Optim. **2**, 435–450 (2007)
268. Wu, C.Z., Teo, K.L., Li, R., Zhao, Y.: Optimal control of switched systems with time delay. Appl. Math. Lett. **19**, 1062–1067 (2006)
269. Xie, G.M., Wang, L.: Controllability and stabilizability of switched linear systems. Syst. Control Lett. **48**, 135–155 (2003)
270. Xiu, Z.L., Song, B.H., Sun, L.H., Zeng, A.P.: Theoretical analysis of effects of metabolic overflow and time delay on the performance and dynamic behavior of a two-stage fermentation process. Biochem. Eng. J. **11**, 101–109 (2002)
271. Xiu, Z.L., Zeng, A.P., An, L.J.: Mathematical modeling of kinetics and research on multiplicity of glycerol bioconversion to 1,3-propanediol. J. Dalian Univ. Technol. **40**, 428–433 (2000)
272. Xu, X.P., Antsaklis, P.J.: Switched systems optimal control formulation and a two stage optimization methodology. In: Proceedings of the 9th Mediterranean Conference on Control and Automation, Dubrovnik, Croatia, June 2001
273. Xu, X.P., Antsaklis, P.J.: Optimal control of switched autonomous systems. In: Proceedings of the 41st IEEE Conference on Decision and Control, Las Vegas, pp. 4401–4406 (2002)
274. Xu, X.P., Antsaklis, P.J.: Results and perspectives on computational methods for optimal control of switched systems. In: Oded, M., Amir, P. (eds.) Hybrid Systems: Computation and Control, pp. 540–555. Springer, Berlin (2003)
275. Xu, X.P., Antsaklis, P.J.: Optimal control of switched systems based on parametrization of the switching instants. IEEE Trans. Autom. Control **49**, 2–16 (2004)
276. Yang, G., Tian, J., Li, J.. Fermentation of 1,3-propanediol by a lactate deficient mutant of *Klebsiella oxytoca* under microaerobic conditions. Appl. Microbiol. Biotechnol. **73**, 1017–1024 (2007)
277. Ye, H., Michel, A.N., Hou, L.: Stability theory for hybrid dynamical systems. IEEE Trans. Autom. Control **43**, 461–474 (1998)
278. Ye, J.X., Zhang, Y.D., Feng, E.M., Xiu, Z.L., Yin, H.C.: Nonlinear hybrid system and parameter identification of microbial fed-batch culture with open loop glycerol input and pH logic control. Appl. Math. Model. **36**, 357–369 (2012)
279. Yong, J.: Systems governed by ordinary differential equations with continuous, switching and impulse controls. Appl. Math. Optim. **20**, 223–235 (1989)
280. Yu, J.B., Xi, L.F., Wang, S.J.: An improved particle swarm optimization for evolving feedforward artificial neural networks. Neural Process Lett. **26**, 217–231 (2007)
281. Zefran, M., Burdick, J.W.: Design of switching controllers for systems with changing dynamics. In: Proceedings of the 37th IEEE Conference on Decision and Control, Tampa, pp. 2113–2118 (1998)
282. Zeng, A.P.: A kinetic model for product formation of microbial and mammalian cells. Biotechnol. Bioeng. **46**, 314–324 (1995)
283. Zeng, A.P., Biebl, H.: Bulk chemicals from biotechnology: the case of 1,3-propanediol production and the new trends. Adv. Biochem. Eng./Biotechnol. **74**, 239–259 (2002)
284. Zeng, A.P., Deckwer, W.D.: A kinetic model for substrate and energy consumption of microbial growth under substrate-sufficient conditions. Biotechnol. Prog. **11**, 71–79 (1995)
285. Zeng, A.P., Ross, A., Biebl, H., Tag, C., Deckwer, W.D.: Multiple product inhibition and growth modeling of *Clostridium butyricum* and *Klebsiella pneumoniae* in glycerol fermentation. Biotechnol. Bioeng. **44**, 902–911 (1994)
286. Zheng, P., Wereath, K., Sun, J., van den Heuvel, J., Zeng, A.P.: Overexpression of genes of the dha regulon and its effects on cell growth, glycerol fermentation to 1,3-propanediol and plasmid stability in *Klebsiella pneumoniae*. Process Biochem. **41**, 2160–2169 (2006)
287. Zheng, Z.M., Cheng, K.K., Hu, Q.L., Liu, H.J., Guo, N.N., Liu, D.: Effect of culture conditions on 3-hydroxypropionaldehyde detoxification in 1,3-propanediol fermentation by *Klebsiella pneumoniae*. Biochem. Eng. J. **39**, 305–310 (2007)
288. Zhu, H., Sun, S.J.: Effect of constant glucose feeding on the production of exopolysaccharides by *Tremella fuciformis* spores. Appl. Biochem. Biotechnol. **152**, 366–371 (2009)